STRUCTURE AND BONDING is issued at irregular intervals, according to the material received. With the acceptance for publication of a manuscript, copyright of all countries is vested exclusively in the publisher. Only papers not previously published elsewhere should be submitted. Likewise, the author guarantees against subsequent publication elsewhere. The text should be as clear and concise as possible, the manuscript written on one side of the paper only. Illustrations should be limited to those actually necessary.

Manuscripts will be accepted by the editors:

Professor Dr. *P. Hemmerich* Universität Konstanz, Abteilung Biologie
D-7750 Konstanz

Professor Dr. *C. K. Jørgensen* 51, Route de Frontenex,
CH-1207 Genève

Professor *J. B. Neilands* University of California, Biochemistry Department
Berkeley, California/USA

Sir *Ronald S. Nyholm,* FRS Professor of Chemistry, University College
Gower Street
London WC 1/Great Britain

Professor Dr. *D. Reinen* Institut für Anorganische Chemie der Universität
Marburg
D-3550 Marburg, Gutenbergstraße 18

Professor *R. J. P. Williams* Wadham College, Inorganic Chemistry Laboratory
Oxford/Great Britain

SPRINGER-VERLAG

D-6900 Heidelberg 1
P. O. Box 1780
Telephone (06221) 491 01
Telex 04-61 723

D-1000 Berlin 33
Heidelberger Platz 3
Telephone (0311) 822001
Telex 01-83319

SPRINGER-VERLAG
NEW YORK INC.

175, Fifth Avenue
New York, N. Y. 10010
Telephone 673-2660

STRUCTURE AND BONDING

Volume 10

Editors: P. Hemmerich, Konstanz
C. K. Jørgensen, Genève · J. B. Neilands, Berkeley
Sir Ronald S. Nyholm, London · D. Reinen, Marburg · R. J. P. Williams, Oxford

With 49 Figures

Springer-Verlag
Berlin Heidelberg GmbH 1972

ISBN 978-3-540-05700-0   ISBN 978-3-540-37027-7 (eBook)
DOI 10.1007/978-3-540-37027-7

Originally published by Springer-Verlag Berlin Heidelberg New York in 1972

Contents

Contents

Kinetics and Mechanism of Alkali Ion Complex Formation in Solution

Dr. Ruthild Winkler

Max-Planck-Institut für biophysikalische Chemie, Göttingen, Germany

Table of Contents

I. Introduction

I.1 Rules for Metal Complex Formation in Solution

Metal complex formation in aqueous solutions is generally a very rapid process. The half time for the rate limiting step, which is considered to be the substitution of a water molecule from the inner coordination sphere of the metal ion, is invariably less than a millisecond. It was not possible to study these reactions in detail until new techniques such as electron spin resonance (ESR), nuclear magnetic resonance (NMR) and relaxation spectrometry were introduced.

During the last decade, characteristic substitution rates for many metal ions have been measured. Fig. 1 (1, 2) gives a condensed survey of some characteristic rate constants. In forming a metal complex, the incoming ligand has to substitute one or several of the water molecules of the inner coordination sphere of the aquo metal complex. The values given in Fig. 1 are related to this (rate limiting) substitution step. Looking at the data, it is reasonable to ask whether there is anything like a

1

"characteristic rate" which, at least with respect to orders of magnitude can be assigned to a given metal ion and is independent of the incoming ligand.

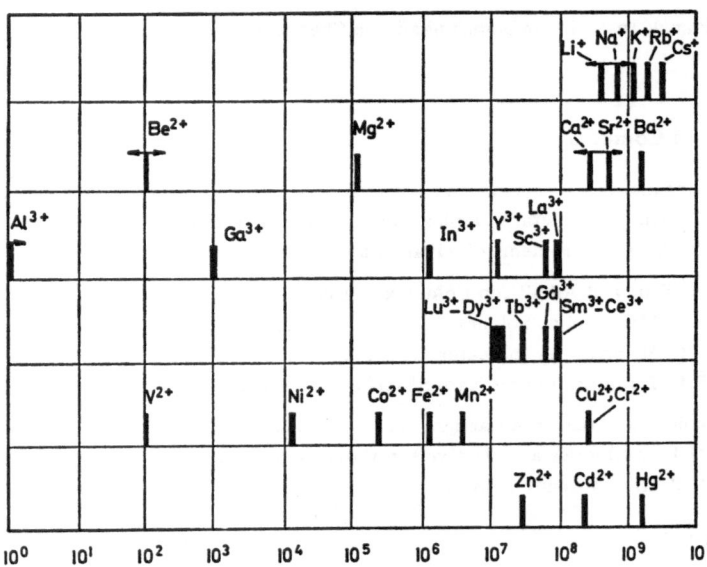

Fig. 1. Characteristic rate constants (sec^{-1}) for substitution of water molecules from the inner coordination sphere of metal ions

Fig. 1 includes three categories of metal ions. In the first group, the water molecules of the inner coordination sphere are so labile that substitution takes place at almost every encounter. The overall rate for the complex formation is thus diffusion controlled. It is not possible to separate one single substitution step from the overall process. The rate constants in this group (to which most of the alkali and some of the alkaline earth ions belong) are therefore only in a very trivial sense characteristic of the nature of the metal ion.

The second group consists of highly charged and small metal ions, e.g. Be^{2+}, Al^{3+}, Fe^{3+} and most of the tetravalent ions. Here the water molecules in the inner coordination shell are so tightly bound that the substitution rate is very low. The electrostatic influence of the metal ion on the water molecule is so strong that internal hydrolysis (i.e. splitting of H_2O into metal-bound OH^- and ligand-bound H^+) occurs prior to substitution. Such a process, of course, is dependent on the basicity of the incoming ligand and is not necessarily expressed in the pH-dependence of the rate constants. As a consequence, the overall rate is largely influenced by the nature of the ligand and thus not characteristic for the metal ion only.

The majority of main group and transition metal ions, however, belong to the third category. These metal ions show specific substitution rates, almost independent of the binding strength of the substituting ligand. Their substitution kinetics may be characterized by three rules:

1. The rate of the substitution of a given solvent molecule in the inner coordination sphere depends mainly on the nature (electron configuration, charge and size) of the metal ion rather than that of the incoming ligand.

2. In multiple substitution, a complexing ligand, which is more tightly bound than the substituted solvent molecule, will *labilize* the rest of the solvent molecules in the inner coordination sphere. On the other hand, if the ligand is less tightly bound it will *stabilize* the other molecules of the inner coordination shell. Here "overall" charge of a multidentate ligand is less important than its "local" charge density of binding strength respectively, which was shown for a number of examples by *Margerum* (*3*), *Hunt* et al. (*4*) and *Hague* (*5*).

3. The substitution rate generally decreases with increasing charge and decreasing size of the metal ion.

Rule 1 and 3 indicate, that the main energy barrier results from loosening the metal ion solvent bond. In a formal sense this process corresponds to a S_N1 type of mechanism. A pure S_N1 mechanism would require a solvent molecule to leave the inner coordination sphere prior to the entry of the new ligand. Here the intermediate state would have a lower coordination number than both the initial and final state and the nature of the entering ligand would then have no significant influence on the rate of substitution. The alternative S_N2 mechanism requires a transition state involving a metal ion with increased coordination number due to significant bond formation between the incoming ligand and the metal ion prior to the departure of the leaving solvent molecule. In this case the rate would be expected to be dependent on the nature of the incoming ligand.

This classification only represents extreme limits. The true mechanism probably resembles a "push-pull" type of substitution process, where the intermediates are not as well defined as in the extremes. The three rules quoted above, which characterize very well the substitution behavior of the third category of metal ions suggest a mechanism close to the S_N1 type in which loosening of the metal ion solvent bond essentially represents the rate limiting step.

Fig. 2 shows the characteristic rate constants for substitution of Ni^{2+} and Cu^{2+} with a variety of different ligands (*2*). Even for multiple

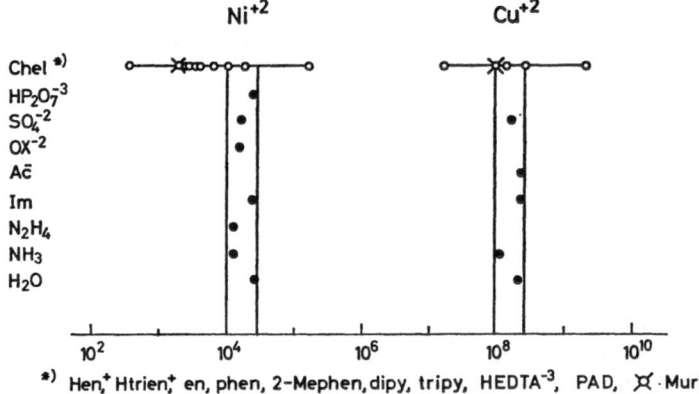

Fig. 2. First and second order rate constants for water substitution of the inner coordination sphere of Ni^{2+} and Cu^{2+} for a series of ligands. The values for simple ligands are corrected to yield the first order rate constants of substitution, whereas overall constants are given for multidentate chelates (values from ref. 2)

substitution processes, in which the first step of substitution is rate limiting (since successive substitutions involving chelation are generally much faster) the rates appear to be nearly independent of the nature of the ligands. Further examples are given in Fig. 3, where the substitution rates of Mg^{2+} and Ca^{2+} are compared (6). For the smaller Mg^{2+} ion (cf. rule 3) the substitution is always about three to four orders of magnitude slower than for the Ca^{2+} ion. (The data listed in Fig. 3 are overall rate constants, which include the charge dependent stability constants of the ion pairs. After correction for the ion pair formation, the substitution rates for Mg^{2+} turn out to be uniformly of the order of magnitude of 10^5 sec^{-1}). The divalent metal ions of the first transition series — as exemplified with Ni^{2+} and Cu^{2+} (Fig. 2) — generally show substitution rates independent of ligand properties. However, as demonstrated by Fig. 4 — these rates do not follow any simple size dependence. This deviation from rule 3 is caused by the dominating influence of the d-electrons. Only d^0-, d^5- and d^{10}-configurations which do not exhibit any ligand field effects, show a monotonic radius dependence comparable to the main group metal ions. Extreme deviations from monotonic behavior are found for the d^3-, d^4-, d^8-, and d^9-configurations, (i.e. V^{2+}, Cr^{2+}, Ni^{2+}, Cu^{2+}). V^{2+} (7) and Ni^{2+} substitute very slowly owing to the strong ligand field stabilization (8) of the symmetrical initial and final state as compared to the unsymmetrical transition states. On the other hand Cr^{2+} (9) and Cu^{2+} are much faster than main group metal ions of comparable charge and size. The Jahn-Teller effect: present in the d^4- and

	$k_{form.}$	
	Mg^{2+}	Ca^{2+}
SO_4^{2-}	$1 \times 10^{5a)}$	
$S_2O_3^{2-}$	$1 \times 10^{5a)}$	
CrO_4^{2-}	$1 \times 10^{5a)}$	$\geqslant 5 \times 10^{7a)}$
F^-	3.7×10^4	
HF	$\sim 4 \times 10^4$	
ATP^{4-}	1.3×10^7	$\geqslant 1 \times 10^9$
$ATPH^{3-}$	$\sim 3 \times 10^6$	
ADP^{3-}	3×10^6	$\geqslant 3 \times 10^8$
$ADPH^{2-}$	1×10^6	
Metal Phthal.	$(\sim 2 \times 10^6)$	$(\sim 7 \times 10^8)$
IDA^{2-}	9×10^5	$2.5 \times 10^{8a)}$
		$Sr^{2+}: 3.5 \times 10^{8a)}$
		$Ba^{2+}: 7 \times 10^{8a)}$
$Glycine^-$		$4 \times 10^{8a)}$
$Oxine^-$	3.8×10^5	$\geqslant 1 \times 10^8$
$Oxine\ H$	$\sim 1 \times 10^4$	

a) first order rate constants

Fig. 3. Rate constants for complex formation reactions of Mg^{2+} and Ca^{2+} (values from ref. 6)

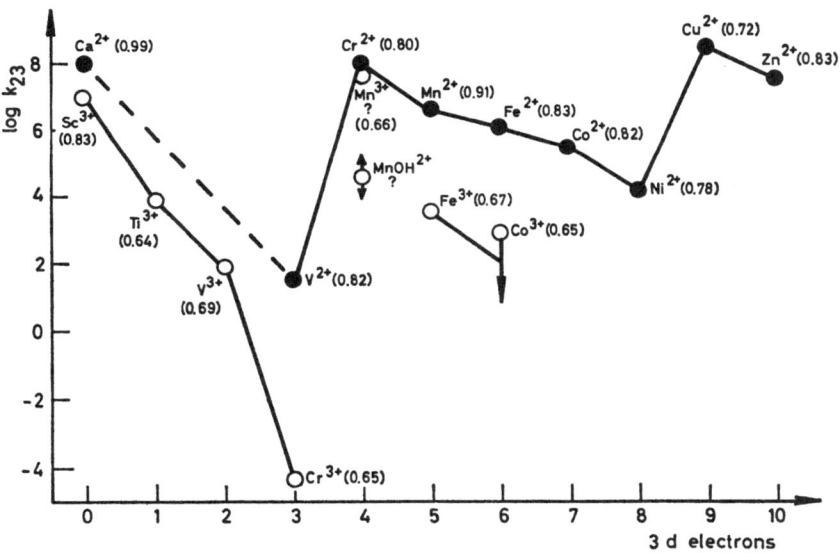

Fig. 4. Characteristic substitution rates (log k_{23}[sec^{-1}]) of di- and trivalent aquo metal ions as a function of the number of 3d-electrons (diagram from ref. 9, *H. Diebler*)

d^9-configuration labilizes the axial positions relative to the very stable planar positions. Here the ligand probably attacks a weakly bonded axial position first, resulting in an unusual fast substitution rate. Since axial and planar positions interchange rapidly chelation can occur very fast via internal conversions.

The substitution behavior of the trivalent d^0-metal ions in the transition earth and rare earth series provides another exception from rule 3. Their rates are even higher than for divalent ions of equivalent size, Sc^{3+} being the most pronounced case (10). A simple monotonic dependence on charge and size can only be expected for ions possessing the same coordination number. Any change in coordination number will greatly influence the binding strength of a given constituent of the inner coordination sphere and thus the rate of substitution. Moreover, an increase of the coordination number beyond six can result in the formation of unsymmetrical complexes. (Only the coordination numbers of four $\simeq$ tetrahedral, six $\simeq$ octahedral and eight $\simeq$ cubic correspond to symmetrical ligand arrangements). Thus labile positions will occur which may be substituted more rapidly than in a symmetrical complex. Fig. 5 contains an illustrative example.

The three rules quoted above are admittedly emprical. They express clearly certain tendencies in the substitution behavior of metal ions.

A precise theoretical approach should be based on an accurate computation of all the single interaction terms, such as: metal ion ligand interactions including electrostatic (ion-ion or ion-multipole and polarization terms) as well as electrostatic and van der Waals terms. Individually

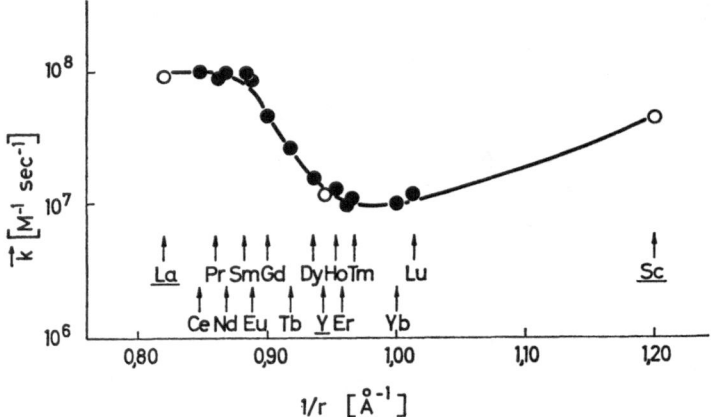

Fig. 5. Second order rate constants for the complex formation of the trivalent lanthanide ions and some d^0-ions of transition metals with murexide as a function of reciprocal ionic radius (values from ref. 10)

these terms are of considerable magnitude, however collectively they tend to compensate each other and contribute only a few kcal/mole to the free energy of complex formation. A fairly exact computation of the single terms would be required in order to yield meaningful values for the equilibrium and kinetic energy parameters of metal complex formation.

I.2 Specific Problems of Alkali Ion Complexes

A consideration of the previous statements leads one to expect a relatively simple substitution "chemistry" for the alkali ions in solution. Due to their "noble gas" like electron configuration, the substitution rates should show a straightforward relationship to physical properties such as charge and size. However, it is naive to assume that complex formation involving main group metal ions is an easily resolved problem. There exist several non-trivial facts, which cannot readily be explained. It is a close examination of these, which will provide some interesting insight into the mechanism of metal complex formation of alkali ions.

1. The free energies of solvation — as well as the solvation enthalpies of alkali ions are as high as about one hundred kilocalories (Fig. 6) (*11*) and yet substitution occurs within 10^{-8} to 10^{-10} seconds (*6*) (cf. Fig. 1).

	$-\Delta H$ (H_2O)		$-\Delta G$ (H_2O)		$-\Delta G_t$ ($H_2O \rightarrow CH_3OH$)
	Eley and *Evans*	*Latimer*	*Eley* and *Evans*	*Latimer*	*Strehlow*
Li^+	133	121	123		
Na^+	115	95	108	91	—0.41
K^+	90	76	85	75	—0.03
Rb^+	81	70	75	70	+0.06
Cs^+	73	62		65	+0.15
Mg^{2+}	501		479		
Ca^{2+}	428		412		
Sr^{2+}	381				
Ba^{2+}	347		334		

Fig. 6. Heats and free energies of solvation of alkali and alkaline earth ions (values from ref. 11)

2. The stability constants of the alkali complexes — and even more so for the far better studied alkaline earth complexes (*12*) — do not show any uniform dependence on the radius of the metal ions (cf. Fig. 7). Any simple model would have to correspond to a monotonic decrease in the

stability constant with increasing metal ion radius or the converse. There are many examples of such types of monotonic radii dependences. However, in the case of chelation with multidentate ligands one often finds maxima and minima in the radius dependence of the stability constants.

	SO_4^{2-}	F^-	Ac^-	IDA^{2-}	NTA^{3-}	$EDTA^{4-}$
Mg^{2+}	2.3	1.8	0.8	3.7	7.0	9.1
Ca^{2+}	2.3	<1.0	0.8	3.4	8.2	11.0
Sr^{2+}			0.4		6.7	8.8
Ba^{2+}		<0.5	0.4	1.7	6.4	7.8
Li^+	0.6				3.3	2.8*
Na^+	0.7				2.2	1.7*
K^+	0.9					

	$DGITA^{4-}$	Metal-Phthal.$^{6-}$	Erio R^{3-}	$Tiron^{4-}$	$Oxine^-$
Mg^{2+}	5.2*	8.9*	7.6*	6.9*	4.7
Ca^{2+}	11.0*	7.8*	5.4*	5.8*	3.3
Sr^{2+}	8.5*			4.6*	2.6
Ba^{2+}	8.4*	6.2*		4.1*	2.1

Fig. 7. Stability constants of alkali and alkaline earth complexes (values from ref. 2). (log K_1 in aq. sol., $\mu \to 0$, * $\mu = 0.1$; $T = 18-25\ ^\circ C$)

3. Many of the main group elements show specific interactions, which are of primary importance in biological systems. There are specific "carriers" which, being highly selective for a given ion size, induce the selective transport of alkali ions through biological membranes. The binding constants of such carriers for Na^+ or K^+ may differ by orders of magnitude.

At first glance these experimental findings would seem to contradict any simple interpretation of the chemistry of the first series of main group metal ions. However, one must admit that there exist only very few experimental data of stability and rate constants on which a detailed discussion can be based. Such studies of thermodynamic and kinetic properties of alkali ion complexes are difficult for three reasons:

a) The complex formation is usually very weak, thus measurements have to be carried out at extremely high metal ion concentrations.

b) According to the high concentrations of reactants required to produce an observable amount of complex, the formation rates fall into the micro- to nanosecond range.

c) These reactions cannot usually be followed spectrophotometrically, because interactions of alkali ions with chromophoric groups are too weak to either produce sufficiently characteristic absorption changes or to compete favourably with protons for binding. In the literature no indicator is described, which forms sufficiently stable complexes with the alkali ions.

Why do some metal ions of a particular size possess extreme stability constants? Is there any simple explanation for size specificity or for a *non*monotonic metal ion radius dependence? Since we are dealing with ions of "noble gas" like electron configuration, the selective complex formation certainly cannot be a consequence of a peculiar chemistry as in the case of transition metal ions. It is more likely to be a special property of the complexing ligand.

II. The Study of Alkali Ion Complex Formation

II.1 Methanol as Solvent

The problems raised in the last section represent the subject of this article, i.e. complex formation of the alkali ions with specific biological carriers, their selectivity and reaction mechanisms. An understanding of the selectivity requires comparative binding studies, and the interpretation of mechanisms might be expected to follow from rate measurements. A correlation of kinetic and thermodynamic data may enable us to understand the principles of carrier action.

In order to solve these problems we have to study both the equilibria and the rates. Since carriers due to their lipophilic character are not soluble in water, all experiments have to be carried out in a nonaqueous but polar medium. Methanol as a solvent offers several advantages for the study of alkali ion complexes:

1. The stability constants for complex formation are invariably higher in methanol than in water, especially if charged ligands are involved. This is due to the lower dielectric constant of methanol. The determinations of stability and rate constants may thus be facilitated for otherwise quite weak complexes, since now either larger observable changes can be induced or lower concentrations can be used, resulting in slower rate processes.

2. The solvation energies of the cation in water and in methanol are very similar as was shown by *Strehlow* (13). The changes in ΔG, ΔH and ΔS values are very small when an alkali ion is transferred from water to methanol. Hence there should not be much difference between the

solvents with respect to ligand substitution reactions so that studies in methanol can also be considered representative for water.

3. Methanol can be used as a solvent for T-jump studies when either the Joule or microwave heating technique is employed. In the case of field effect and soundabsorption measurements charged ligands even show larger effects (owing to the lower dielectric constant of the solvent). For uncharged ligands (e.g. the macrotetrolides) relatively large ΔH but comparatively small ΔV values are to be expected. This means that the equilibrium will be strongly temperature dependent. Sound waves in methanol cause a pronounced temperature effect, which in water around room-temperature is almost absent (due to the density maximum at 4 °C).

Binding and rate studies of alkali ion carriers in methanol require a suitable indicator, if spectrophotometric observation is used. This is because the complex formation of the alkali ion with the carrier does not involve any easily observable absorption change. Such an indicator has to fulfill several requirements:

1. The indicator should be soluble in methanol.

2. The absorption spectrum should change detectably upon addition of alkali ions, in a wavelength range convenient for relaxation methods.

3. The reaction of alkali metal ions with indicator should be rapid compared with the reaction of the metal ion with carrier.

4. The indicator should be applicable in the concentration range of 10^{-3} to 10^{-5} moles of alkali ions per liter, since the dissociation constant of carrier complex formation is to be expected in this range.

5. There should be no interaction between the indicator and the carrier which might affect the binding of the metal ion to the carrier.

II.2 Murexide as Indicator

Murexide, the ammonium salt of purpuric acid, was known for long to be a good indicator for Ca^{2+} in aqueous solution (14). The chemical composition of the dye is as follows:

At neutral pH it is present as a monovalent anion, the negative charge being distributed among the four oxygens adjacent to the N-bridge. Protonation of the anion only occurs at relatively low pH. The corresponding pK-value in water is around 0. In methanol it shifts to about 4.5. Since the alkali ions generally form quite weak complexes it is very important that protons do not compete too strongly for the binding site. At alkaline pH two of the four protons of the NH-groups can be removed. In water these pK-values are 9.2 and 10.5 and correspondingly higher in methanol.

Murexide forms chelates with all the alkali and alkaline earth ions, but the strength of the complexes varies considerably. The chelation involves the N-bridge as well as the neighbouring oxygens. All complexes are distinguished by strong spectral shifts as compared to the free anion.

The spectral behaviour of murexide complexes is quite unique, consisting of strong and ion specific absorption shifts towards shorter wavelengths. In general electrostatic interactions of a bound metal ion do not produce such a pronounced effect on the electronic structure and hence the spectrum of a dye. Twisting of the two rings relative to each other, brought about by complexation, seems to offer a better explanation for the observed spectral properties of murexide complexes.

II.3 Equilibrium and Rate Studies

Complex formation of the alkali ions with murexide in methanol was studied quantitatively by spectrophotometric titration with Li^+, Na^+, and K^+. (For Rb^+ and Cs^+ only qualitative measurements could be obtained since these complexes tend to precipitate). Fig. 8 shows the shift of the absorption maximum upon titration with Na^+. The well defined isosbestic point is a good indication for a simple $1:1$ complexation equilibrium. In so much as the spectral shift (upon complexation) is a criterion of the strength of the complexes, Fig. 9 indicates that the "absolute" complex stability parallels monotonically the sequence of ionic sizes. (Both $\Delta\lambda_{max}$ and $\Delta\varepsilon$ are largest for the smallest ion). In the alkali ion series Li^+ forms the strongest and Cs^+ the weakest complexes. This monotonic size dependence of the charge density is also expressed in the energy values for the desolvation ($-\Delta H_{hydr.}$ for $Li^+ = 120$ kcal and for $Cs^+ \simeq 60$ kcal) (11).

Apparently in contrast to this simple interpretation is the non-monotonic behaviour of the thermodynamic stability constants for Li^+, Na^+, K^+, Rb^+ and Cs^+; Na^+ possessing the highest value. However, one has to remember that thermodynamic stability constants are not a measure of the absolute strength of a complex, but rather of the relative strength as compared to solvation. Thus the maximum for Na^+ results

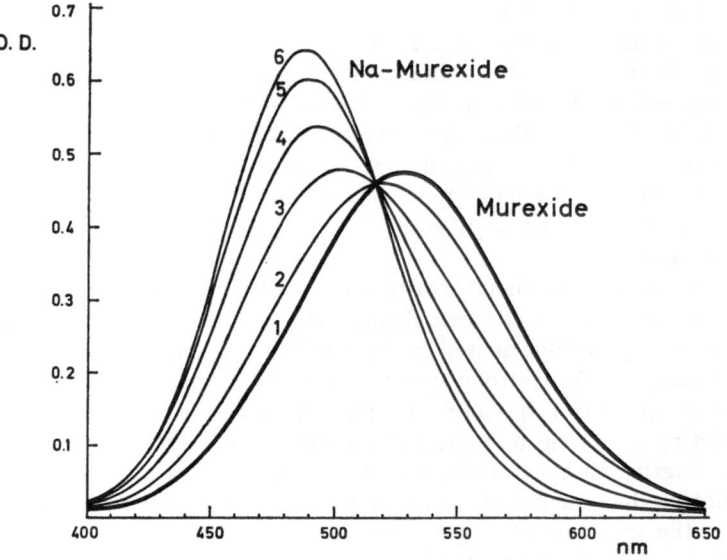

Fig. 8. Spectrophotometric titration of murexide with Na^+ (25 °C; $c_{Mur}^0 = 4 \times 10^{-5}M$)

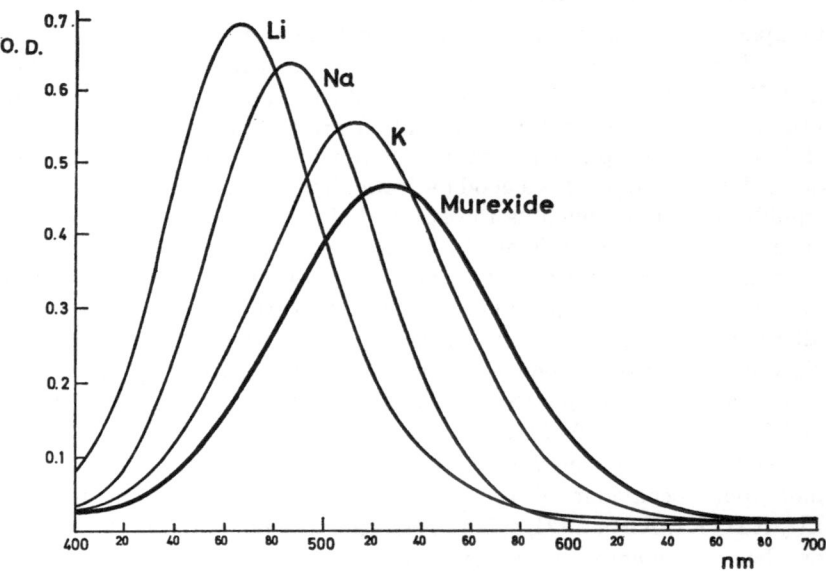

Fig. 9. Absorption spectra of alkali metal ion complexes with murexide

from a balance between binding and solvation energy. Na^+ yields the largest gain in energy upon complexation compared to both Li^+ and Cs^+. The explanation of this peculiar effect already contains the essential feature for the explanation of selectivity.

If murexide is to be used as an indicator in studies of fast processes it is important that not only pronounced spectral shifts are present, but also that they occur rapidly compared to the reaction under study. The indicator may still be used if its rate of complex formation is comparable to that to be followed. In this situation, however, it is necessary to know accurately the rate constants of the indicator reaction for the evaluation of the resulting relaxation spectrum of the coupled reactions.

The relaxation times for the complex formation of Li^+ and Na^+ with murexide have been determined. In these studies the electric field pulse technique was used, which has a resolution time of about 30 nanoseconds. The method is described in detail elsewhere (15). Fig. 10 shows a typical relaxation curve for the Na-murexide reaction. The effect for K-murexide complex formation (requiring higher concentration, due to the lower stability constant) was already beyond the resolution of this method.

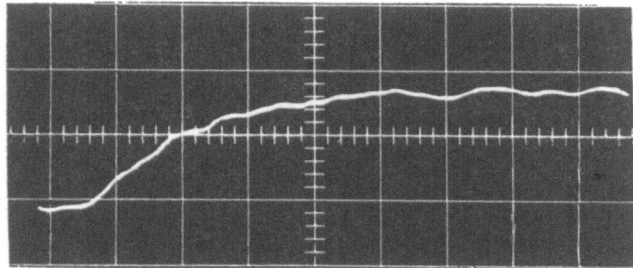

Fig. 10. Oscillogram of field effect relaxation for the Na-murexide system in methanol. (Horizontal time scale: 100 nsec/large division)

However, one may derive a lower limit for the rate constant which almost coincides with the upper limit for diffusion controlled processes. Even the measured value for Na^+ is already very close to this limiting value of the rate constant of about 2 to $3 \times 10^{10} M^{-1} sec^{-1}$. (For charged reaction partners this value is somewhat higher in methanol than in water due to the lower dielectric constant of methanol favouring the electrostatic attraction of the reactants). These results indicate that the rate constants for K^+, Rb^+ and Cs^+ should be diffusion controlled. Only

Li$^+$ shows a considerably lower value. The rate and equilibrium constants of the complex formation of alkali ions with murexide in methanol are summarized in Fig. 11. These data clearly point out that murexide — both from the static and dynamic point of view — is an ideal candidate for the indication of alkali ions in methanol. The spectral shifts are easily detectable and characteristic. The stability constants are in a very convenient range (especially with regard to an investigation of the complex formation of the carriers). The rate of the complex formation is very high so that any change in the coupled reaction system can be followed almost instantaneously.

25 °C	k_{form} [M$^{-1}\cdot$ sec^{-1}]	k_{diss} [sec^{-1}]	K [M^{-1}]
Li$^+$	$5.5\cdot 10^9$	$7.7\cdot 10^6$	$7.1\ \cdot 10^2$
Na$^+$	$1.5\cdot 10^{10}$	$5.9\cdot 10^6$	$2.55\cdot 10^3$
K$^+$	$\sim 2\cdot 10^{10}$ a)	$> 10^7$	$1.1\ \cdot 10^3$

a) diffusioncontrolled

Fig. 11. Stability and rate constants for alkali metal ion murexide complexes in methanol

Analogous measurements were carried out for the alkaline earth complexes with murexide. Generally, the stability constants are considerably higher than those of the alkali metal complexes. The constants in methanol are as much as six orders of magnitude greater than those in water. This is due to the stronger contribution of the electrostatic interactions between the negatively charged dye molecule and divalent ions in a solvent of relatively low dielectric constant.

Only in the case of Mg^{2+} was simple 1:1 complex formation found, whereas Ca^{2+}, Sr^{2+} and Ba^{2+} have a clear tendency to form 1:2 complexes. For Mg^{2+} the binding of the second ligand is so weak that it could be neglected. Equilibrium studies yielded a value for K_1 (the stability constant for the binding of the first ligand) of 5.5×10^5M^{-1}. From spectroscopic measurements with Ca^{2+} and Sr^{2+} only the second step i.e. K_2 could be detected. The values are: $K_2 \simeq 5 \times 10^5$M^{-1} for Ca^{2+} and $K_2 \simeq 2 \times 10^5$M^{-1} for Sr^{2+}. An estimate of K_1 for Ca^{2+} may amount as high as 10^9M^{-1}. The relative variation in the stability constants is very similar to that obtained in water (16). Thus Ca^{2+} forms the most stable complexes. Its ionic radius (0.99 Å) corresponds to that of Na$^+$ (0.98 Å) in the alkali ion series, which also shows a maximum of stability with murexide. The absolute binding strength of the alkaline earth complexes with murexide however decreases monotonically with increasing ionsize (Fig. 12).

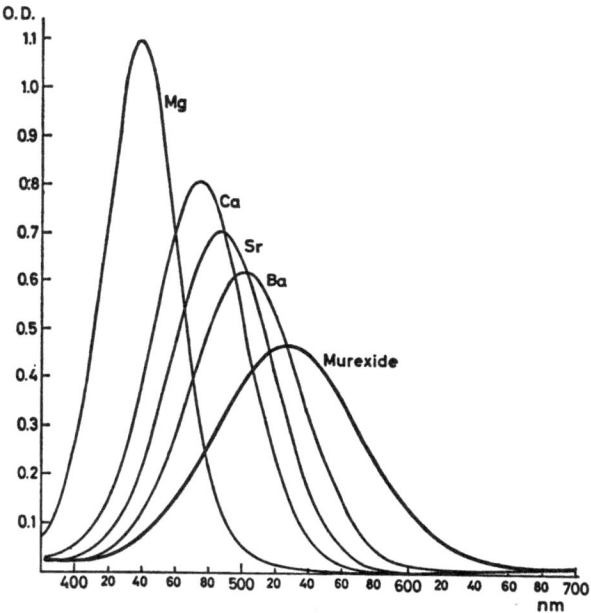

Fig. 12. Absorption spectra for the complexes of alkaline earth ions with murexide

The substitution rate for Mg^{2+} was measured by using the temperature-jump technique. The value obtained for the overall rate constant (k_{form}) is $8 \times 10^6 M^{-1}sec^{-1}$. The absolute values of the rate constants in methanol are again only slightly higher as compared to water (cf. Fig. 3) as a consequence of the more stable ionpair formation in methanol ($DK = 31.5$). Taking into account the stability of the ionpair, one obtains a substitution rate for Mg^{2+} in the range of $10^5 sec^{-1}$, for Ca^{2+} around $5 \times 10^8 sec^{-1}$ and for Sr^{2+} and Ba^{2+} even higher. These data may be compared to those for water as a solvent, even though for Mg^{2+} they correspond to the first complex formation step, and for Ca^{2+}, Sr^{2+} and Ba^{2+} to the second one. Measurements with Mg^{2+} in water show that successive substitutions occur at comparable rates.

This section not intended to be a comprehensive survey of the numerous studies on complex formation of alkaline earth ions, but rather to demonstrate the usefulness of murexide as an indicator in methanol and especially the similarity of substitution rates for both methanol and water as a solvent (18).

III. Biological Alkali Ion Carriers

III.1 Antibiotics as Selective Ion Carriers

Some years ago *Moore and Pressman* (19) as well as *Lardy* (20) and coworkers found that certain antibiotics — isolated from microorganisms — facilitate specifically the transport of alkali ions through membranes. These "carriers" are characterized by a selectivity, which enables them to distinguish between two chemically so similar ions such as K^+ and Na^+, by orders of magnitude. Their chemical structure conveniently separates them into two groups: one consisting of macrocyclic and the other of open chain structures. The compounds of the latter class, e.g. nigericin etc. possess a carboxyl group, which is dissociated at neutral pH range. One could easily imagine that this negatively charged group acts as a ligand which stabilizes the carrier complex via its electrostatic interaction with the positively charged metal ion. However, it turned out, that this carboxyl group is not directly involved in the complex formation. X-ray studies indicate that this group is used to form hydrogenbonds with the two terminal hydroxyl groups (at the other end of the chain molecule), resulting in a cyclic configuration (21). It is now possible for all the (polar) furan oxygens to participate in the complex formation. Cyclic structures generally show some advantage as compared to linear ones, especially in this case of complex formation with multidentate ligands. Actually, most of the carrier compounds which exhibit significant selectivity have a macrocyclic configuration.

The principal representatives of the antibiotics with carrier properties are listed in Fig. 13. The macrocyclic species can be peptides (as

---

Macrocyclic Compounds:

Valinomycin
Gramicidin S, Tyrocidin A
Enniatins
Macrotetrolides
Antamanide
(Gramicidin A, B, C: acyclic)
Cyclic Polyethers (Crown compounds)

Open-chain Compounds:

Nigericin
Monensin
Grisorixin
(Dianemycin)

Fig. 13. Ion selective antibiotics

gramicidin and antamanide) depsipeptides (as valinomycin and the enniatins) or macrolides (as nonactin and its homologues). In addition, a whole series of cyclic polyethers synthetized in the laboratory (cf. *Pedersen*) have many carrier properties in common with the natural antibiotics. The mechanisms for these compounds have been studied recently by *P. B. Chock* (*36*).

The cyclic substances so far investigated may in short be characterized by the following properties (*22*):

1. at physiological pH ($pH = 7$) they are electrically neutral
2. they induce the transport of alkali metal ions across both natural and artificial lipid membranes
3. most of them show a pronounced K^+ over Na^+ selectivity.

The members of the second category (open-chain compounds)

1. possess a negative charge at physiological pH — due to the dissociated carboxyl group at pH 7, whereas
2. nigericin shows a clear K^+ over Na^+ selectivity, monensin favourably binds Na^+ and dianemycin only slightly prefers Na^+ over K^+.

Rate studies have been carried out with some of these carriers and in particular for the macrotetrolides, which are considered in more detail below.

The macrotetrolides (also called actins) can be isolated from actinomycetes (*23*) as was shown by Prelog and his school. Their chemical composition according to *Gerlach and Prelog* (*24*) is depicted in Fig. 14. The homologues of nonactin: monactin, dinactin and trinactin differ by

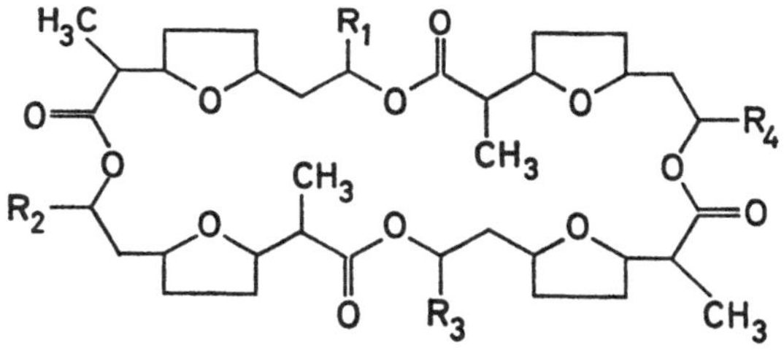

$R_1 = R_2 = R_3 = R_4 = CH_3$ Nonactin

$R_1 = R_2 = R_3 = CH_3$ $R_4 = C_2H_5$ Monactin

$R_1 = R_3 = CH_3$ $R_2 = R_4 = C_2H_5$ Dinactin

$R_1 = CH_3$ $R_2 = R_3 = R_4 = C_2H_5$ Trinactin

Fig. 14. Chemical structure of macrotetrolides

the presence of additional methyl groups in the positions R_2, R_3 and R_4. With the help of X-ray analysis (25) of a K^+-nonactin crystal, *Dunitz et al.* were able to determin the spatial configuration. In the complex, the cyclic backbone of the actin molecule is twisted like the seam of a tennis ball as is drawn schematically in Fig. 15. This particular spatial arrangement allows the carrier to surround and enclose the alkali ion completely. All polar groups — four furan and four keto oxygens — point towards the center, thus building up a quasi cubic structure around the cation. The nonpolar groups are situated on the periphery of the molecule. In the closed form the actin molecule appears to be a hydrophobic sphere. This unique distribution of the polar and nonpolar groups accounts for the special properties of the carriers: their selectivity for certain ions and their solubility in nonpolar media such as lipid membranes. Equilibrium studies of Na^+ and K^+ complexes with nonactin and monactin show a strong preference of K^+ as shown by *Simon* and coworkers (26). The binding constants for K^+ with the various macrotetrolides increase steadily from nonactin to the higher homologues.

The highest specificity for K^+ has so far been found for valinomycin. The data from conductivity measurements by *Shemyakin* and his school (27) and also spectrophotometric titration measurements in the U. V. by *Funck and Grell* (28) show that the K^+ complex is about 2000 times more stable than the corresponding Na^+ complex. The chemical composition of valinomycin is given in Fig. 16. The spatial conformation

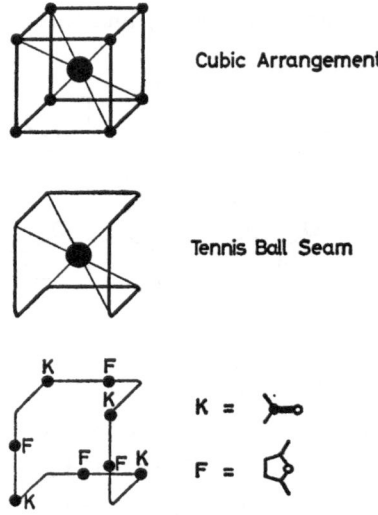

Cubic Arrangement

Tennis Ball Seam

K =

F =

Fig. 15. Schematic representation of the spatial configuration of alkali metal ion macrotetrolide complexes

Valinomycin

D-Hy-i-Valac   D-Val   L-Lac   L-Val

$$\left[\begin{array}{c} \overset{\displaystyle \underset{\displaystyle CH}{\overset{\displaystyle H_3C \quad CH_3}{\diagdown\diagup}}}{\underset{\displaystyle O}{|}} \end{array}\right]$$

```
 H3C   CH3  H3C   CH3                H3C   CH3
   \   /      \   /                    \   /
    CH         CH          CH3          CH
    |          |           |            |
 —O—CH—C—NH—CH—C—O—CH—C—NH—CH—C—
         ||         ||        ||           ||
         O          O         O            O
```
  $_3$

Fig. 16. Chemical structure of valinomycin

of the complex (*29, 30*) — analysed independently by two different groups — (*Shemyakin et al., Steinrauf et al.*) is an octahedron. Here again — in analogy to the actincomplexes — all the polar groups are arranged around the metal ion in the center of the molecule, whereas the nonpolar groups are pointing to the outside. A peculiarity of valinomycin is that it can form six hydrogenbonds, (and thereby change its conformation in discrete steps). The complex and the positions of these six hydrogen-bonds are pictured schematically in Fig. 17. It can be shown that internal structural changes play an important role during the complex formation between the carrier and the metal ion.

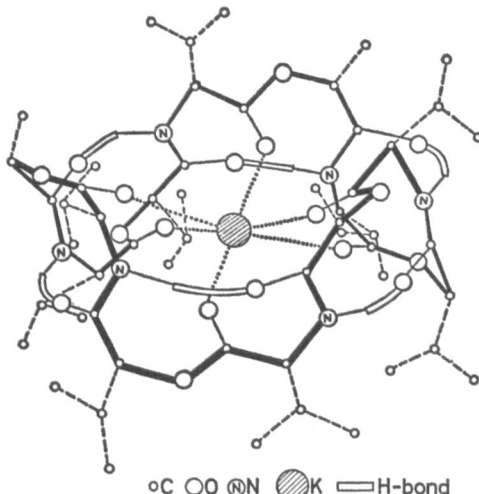

○C ○O ⓝN ⬤K ▭H-bond

Fig. 17. Conformation of the K$^+$ complex of valinomycin. (reproduced from *M. M. Shemyakin et al.* ref. 38)

III.2 Rate Measurements

Detailed information about the mechanism of the carrier complex formation can be obtained from kinetic measurements. Various relaxation techniques have been applied to both equilibrium and rate studies with biological alkali ion carrier complexes. These methods were described in detail in several review articles: cf. ref. 31.

Since the relaxation times of the carrier complexes appear to be distributed over a wide time range they can be determined only by using a variety of different relaxation devices.

Temperature-jump (31) measurements (with a time resolution in the microsecond range) only yielded a lower limiting value for the rate of the monactin-Na$^+$ system. The relaxational amplitudes of these measurements could be evaluated in order to obtain the stability constant (K_{stab}) and heat of reaction (ΔH) (18). The time constant for the reaction of monactin with Na$^+$ was obtained by the sound absorption technique (32). The rate constant of recombination was found to be (k_{form}) $2-3 \times 10^8 M^{-1}$ sec^{-1}. The complex formation of trinactin with Na$^+$ (at high ionic strength) was just detectable with the T-jump technique since the time constant for this system was sufficiently larger than the time constant of heating (33).

Another compound studied was antamanide. It is a cyclic decapeptide isolated from amanita phalloides. Its structure was determined by Th. Wieland and his coworkers (34). This compound was also found to have carrier properties, however with a pronounced Na$^+$ specificity. The relaxation time of the complex formation process between antamanide and Na$^+$ is in the millisecond range (35). It could therefore be investigated by means of the T-jump technique.

The high time resolution of the electric field pulse device turned out to be of advantage for the study of nigericin (36). Unfortunately, this technique could not be applied to the electrically neutral macrocyclic compounds, since it is based on the presence of a dissociation field effect. However, nigericin and murexide do give a dissociation field effect. The rate of complex formation between Na$^+$ and nigericin was found to be as high as that for murexide, i.e. diffusion controlled.

The rates of complex formation of valinomycin with Na$^+$ and K$^+$ were obtained by a combination of measurements carried out with the help of the sound absorption and T-jump method. Both techniques were required in order to resolve the relaxation spectrum, which includes the rate processes for complex formation as well as for structural changes of the valinomycin molecule (37).

The values of the stability and rate constant for all the carrier compounds with Na$^+$ studied so far are listed in Fig. 18. For comparison,

Carrier	$k_{form}[M^{-1}sec^{-1}]$	$k_{diss}[wec^{-1}]$	$K_{stab}[M^{-1}]$	$\Delta H[kcal/mole]$	Ref.
Murexide	1.5×10^{10}	5.9×10^6	2.5×10^3	$+1.4$	(18)
Nigericin	$\sim 2 \times 10^{10}$	$\sim 2 \times 10^6$	1×10^4	$+2.3$	(36)
Monactin	$\sim 2 \times 10^8$	4×10^5	5×10^2	-6.0	(18)
Dinactin	6×10^7	5×10^4	1×10^3	-6.6	(33)
Trinactin	7×10^7	4×10^4	2×10^3	-7.3	(33)
Valinomycin	7×10^6	5×10^5	13	$-$	(37)

Fig. 18. Rate constants, stability constants and heats of reaction for the complex formation of Na^+ with carriers in methanol

the data for the rate and stability constant of murexide are included in Fig. 18. In the first part of this article it was shown that the complex formation of murexide with Na^+ is an almost diffusion controlled process with a rate constant of $1.5 \times 10^{10}M^{-1}sec^{-1}$. In a very trivial sense one also may call murexide a carrier, however with Na^+ specificity (similar to antamanide and some of the carriers of the so called nigericin-group). Its stability constant for Na^+ is about two times higher than for any of its neighbouring cations (namely Li^+ and K^+). The positive value of the enthalpy of reaction (ΔH) indicates the predominant electrostatic influence reflecting the temperature dependence of the dielectric constant. One always expects such a positive value for electrostatic interactions.

The behaviour of the nigericin-Na^+ system is very similar. Here again an extremely high rate for the recombination of nigericin with Na^+ is found and a positive ΔH value, indicating the electrostatic nature of the interaction.

The macrocyclic compounds seem to react more slowly. One has, however, to take into account that these molecules are uncharged. The limiting value of a diffusion controlled reaction with an uncharged species is expected to be only $3 \times 10^9 M^{-1}sec^{-1}$. This means that for the monactin, dinactin and trinactin the reaction rates are about 10 to 50 times slower than for a diffusion controlled process. (The numbers are approximate, since measurements of these very high rates turn out to be quite difficult). Nevertheless, the high rate of complex formation is surprising since, during the encounter between the carrier and the solvated metal ion in solution all the solvent molecules of the inner coordination sphere of the metal ion have to be substituted by the ligands of the carrier. Assuming that there is a 50% probability for a stepwise exchange of every single solvent molecule during the encounter, then such a single step of substitution occurs within 10^{-9} seconds. A time constant of similar order of magnitude has already been found for substitution reactions of the alkali metal ions with simple ligands as shown in Fig. 1. In contrast to the charged chelate ligand the macrotetrolides show negative values for the heat of reaction. The value of ΔH indicates that there must be a considerable entropy increment due to interaction with solvent molecules involved in this process. There are even indications for a true two step mechanism, if the rate data are analyzed in detail.

The results of the kinetic studies with valinomycin (37) are of special interest. In this case the reaction is about one order of magnitude slower than for the actins. This fact coincides with the rates, which have been found for the structural changes in this molecule. The conformation of valinomycin is stabilized by internal hydrogen bonds. Sound absorption measurements of valinomycin in methanol yielded three relaxation times in the time range of 10^{-9} to 10^{-7} seconds. Apparently the slowest structural change is rate limiting for the loading of valinomycin with the metal ion.

From this information on rate and stability constants of the various carrier complexes we may derive four rules for the "design" of an optimal carrier:

1. The architecture of a carrier molecule must be such that the polar ligands are buried in the interior of the molecule, whereas the surface is coated by hydrophobic groups. This gives the carrier molecule a high solubility in lipid phases.

2. The polar ligands must be able to compete favourably with the solvent molecules bound in the inner coordination sphere of the metal ion. Their number should be close to the coordination number of the metal ion. Specificity results from the difference of solvation energies of the various metal ions.

3. The polar groups inside the carrier molecule should be arranged in such a way that they can form a cavity of optimal size adapted to the metal ion to be selected. This is best achieved by chelate like arrangement of the ligands with a defined stereochemistry with respect to each other and to the metal ion. Optimal fit does not necessarily occur at minimum size of the cavity; it is rather related to steric fixation of the ligands where the differences of the free energies of ligand binding and solvation is maximal. Cavity formation will involve both ligand-ligand repulsion as well as steric hindrance of the ligands in the multidentate chelate.

4. The structure must be flexible in order to allow the metal ion to penetrate as quickly as possible into the cavity, which is defined by the special arrangements of the ligands. Such a flexibility is most easily achieved by rapid conformation changes involving a switching from an open to a closed form and vice versa.

The mechanism of the carrier complex formation must involve stepwise substitution of solvent molecules of the inner coordination sphere of the metal ion. In each step the energy of desolvation must be compensated for by the ligand binding energy so that the total binding energy is conserved during the whole loading process. This requires the solvated metal ion to come as closely as possible in contact with the polar groups of the carrier. A rigid arrangement of the ligands of the carrier would not allow such a rapid substitution. Therefore a fast conformation change seems to be an important prerequisite of carrier action, allowing an optimal balance between high selectivity (i.e. high binding strength) and fast performance (i.e. high rate of complex formation). These two properties usually exclude each other.

' High rates are actually necessary for an efficient carrier action. A metal ion to be transported across a membrane must be accepted and delivered by the carrier as quickly as possible. The specificity is defined by the difference in stability constants of complex formation for different metal ions. The stability constant on the other hand determines the maximal values of rate for the delivery of the metal ion. If this rate of delivery would become slower than the actual transport rate all advantages gained by higher selectivity (high stability constant) would be lost (due to the slower rate of delivery). This condition requires the rate of complex formation to be almost diffusion controlled which can only be facilitated via a conformation change of the carrier molecule.

Acknowledgement is made to *Roger Thornley* for valuable suggestions.

I am very much indebted to *Manfred Eigen* for his steady support and criticism in the preparation of the manuscript.

References

1. *Eigen, M.:* Pure Appl. Chem. 6, 97 (1963).
2. *Eigen, M., Wilkins, R. G.:* In Mechanism of Inorganic Reactions, Advan. Chem. Ser. 49, 55 (1965).
3. *Margerum, D. W., Rosen, H. M.:* J. Am. Chem. Soc. 89, 1088 (1967).
4. *Marks, S., Dodgen, H. W., Hunt, J. P.:* Inorg. Chem. 7, 836 (1968).
5. *Hague, D. N., Eigen, M.:* Trans. Faraday Soc. 62, 1236 (1966).
6. *Diebler, H., Eigen, M., Ilgenfritz, G., Maass, G., Winkler, R.:* Pure Appl. Chem. 20, 93 (1969).
7. *Eigen, M.:* Ber. Bunsenges. physik. Chem. 67, 763 (1963).
8. *Basolo, F., Pearson, R. G.:* Mechanisms of Inorganic Reactions, 2nd ed., John Wiley, N. Y. (1967).
9. *Diebler, H.:* Ber. Bunsenges. physik. Chem. 74, 268 (1970) — *Kruse, W., Thusius, D.:* Inorg. Chem. 7, 464, (1968).
10. *Geier, G.:* Ber. Bunsenges. physik. Chem. 69, 617 (1965).
11. a) *Eley, D. D., Evans, M. G.:* Trans Faraday Soc. 34, 1093 (1938).
    b) *Latimer, W. M., Pitzer, K. S., Slansky, C. M.:* J. Chem. Phys. 7, 108 (1939).
    c) *Strehlow, H.:* Z. Elektrochem., Ber. Bunsenges. physik. Chem. 56, 119 (1952).

12. *Martell, A., Sillén, L. G.:* Stability Constants of Metal Ion Complexes, The Chemical Society, London (1964).
13. *Strehlow, H.:* In Chemistry of Non-Aqueous Solvents, Vol. 1, p. 129, Academic Press N. Y. (1966).
14. *Schwarzenbach, G., Gysling, H.:* Helv. Chim. Acta, *32* 1314 (1949). — cf. also *Schwarzenbach, G., Flaschka, H.:* Die komplexometrische Titration, 5. Aufl., Ferdinand Enke, Stuttgart (1965).
15. *Ilgenfritz, G.:* Dissertation, Göttingen (1966).
16. *Geier, G.:* Helv. Chim. Acta, *50*, 1879 (1967).
17. *Hague, D. N., Eigen, M.:* Trans. Farad. Soc. *62*, 1236 (1966).
18. *Winkler, R.:* Dissertation, Wien—Göttingen (1969).
19. *Moore, C., Pressman, B. C.:* Biochem. Biophys. Res. Comm. *15*, 562 (1964).
20. *Graven, S. N., Lardy, H. A., Johnson, D., Rutter, A.:* Biochemistry *5*, 1729 (1966).
21. *Steinrauf, L. K., Pinkerton, M., Chamberlin, J. W.:* Biochem. Biophys. Res. Comm. *33*, 29 (1968).
22. *Simon, W.:* Neurosciences Res. Prog. Bull., Vol. 9, No. 3, p. 308.
23. *Corbaz, R., Ettlinger, L., Gäumann, E., Keller-Schierlein, W., Kradolfer, F., Neipp, L., Prelog, V., Zähner, H.:* Helv. Chim. Acta *38*, 1445 (1955).
24. *Gerlach, H., Prelog, V.:* Liebigs Ann. Chem. *669*, 121 (1963).
25. *Kilbourn, B. T., Dunitz, J. D., Pioda, L. A. R., Simon, W.:* J. Mol. Biol. *30*, 553 (1967).
26. *Pioda, L. A. R., Wachter, H. A., Dohner, R. E., Simon, W.:* Helv. Chim. Acta *50*, 1373 (1967).
27. *Shemyakin, M. M., Ovchinnikov, Yu. A., Ivanov, V. T., Antonov, V. K., Shkrob, A. M., Mikhaleva, I. I., Evstratov, A. V., Malenkov, G. G.:* Biochem. Biophys. Res. Comm. *29*, 834 (1967).
28. *Eggers, F., Funck, Th., Grell, E.:* publication in preparation.
29. *Ivanov, V. T., Lain, I. A., Abdulaev, N. D., Senyavina, L. B., Popov, E. M., Ovchinnikov, Yu. A., Shemyakin, M. M.:* Biochem. Biophys. Res. Comm. *34*, 803 (1969).
30. *Pinkerton, M., Steinrauf, L. K., Dawkins, Ph.:* Biochem. Biophys. Res. Comm. *35*, 512 (1969).
31. *De Maeyer, L.:* Ber. Bunsenges. physik. Chem. *64*, 65 and 80 (1960).
    De Maeyer, L., Kustin, K: In Annual Review of Physical Chemistry, *14*, 5 (1963).
    Eigen, M., De Maeyer, L.: In Technique of Organic Chemistry, A. Weissberger, Ed., 2nd ed. Vol. VIII/2, Interscience, N. Y. (1963).
    Kustin, K.: "Methods in Enzymology", Vol. XVI Fast Reactions, Academic Press N. Y. and London (1969).
32. *Eggers, F.:* Acustica, *19*, 323 (1967/68).
33. *Chock, P. B., Eggers, F., Eigen, M., Winkler, R.:* publication in preparation.
34. *Wieland, Th., Lüben, G., Ottenheym, H., Faesel, J., De Vries, J. X., Konz, W., Prox, A., Schmid, J.:* Angew. Chem. *80*, 209 (1968).
35. *Burgermeister, W., Winkler, R.:* work in progress.
36. *Chock, P. B.:* publication in preparation.
37. *Eggers, F., Funck, Th., Grell, E.:* publication in preparation.
38. *Shemyakin, M. M., Ovchinnikov, Yu. A., Ivanov, V. T., Antonov, V. K., Vinogradova, E. I., Shkrob, A. M., Malenkov, G. G., Evstratov, A. V., Laine, I. A., Melnik, E. I., Ryabova, I. D.:* J. Membrane Biol. *1*, 402 (1969).

Received March 22, 1971

Intra- and Inter-Molecular Bonding and Structure of Inorganic Pseudohalides with Triatomic Groupings

Z. Iqbal

Explosives Laboratory, FRL, Picatinny Arsenal, Dover, N. J. 07801, U.S.A.

Table of Contents

I. Introduction

The term "pseudohalide" was first introduced by *Birkenbach* and *Kellerman* (1) to describe polyatomic groups resembling halides in their chemical properties and will be used here in this context to describe the inorganic derivatives of the sixteen valence electron triatomic ions: N_3^- (azide), CNO^- (fulminate), NCO^- (cyanate) and NCS^- (thiocyanate). This work will be concerned with a correlative review of the structural physics and chemistry of the relatively simple metal pseudohalides and their hydracids. The discussion will be divided into sections on molecular and crystal geometry, vibrational energy levels, bonding forces and electronic energy levels in these compounds. The aim will be to bring together the available

25

structural information and particularly to pinpoint interesting intra- and inter-group comparisons. In the concluding section the correlation of structure with the thermal stability of these compounds will be briefly explored in order to focus attention on the importance of structural studies in understanding the exotic thermal behaviour of the azides and fulminates in particular.

For the purposes of this paper it will be assumed that the reader is familiar with the elements of molecular and crystal structure and the application of group theory to structural problems. No attempt will be made to describe experimental methods and discussion of established ideas will be introduced only for the sake of continuity in presenting an argument. Although in a few instances previous data will be re-interpreted and some new simplified generalizations made, this paper will not introduce new experimental and theoretical results. The discussions will be broadly arranged in order of increasing molecular complexity, with the crystalline compounds receiving the most of attention.

II. Molecular and Lattice Geometry

The basic pseudohalide unit is a linear triatomic group which exists as an anion in polar and semi-polar crystals. Inter-group differences in ionicity among the solids are a function of the electron affinity of the anions which increase in the order $NCS^- < N_3^- < CNO^- < NCO^-$ (2), while intra-group variations depend to a large degree on the ionization potentials of the metal ions. The structural parameters which will be considered below together with a discussion of phase transformations have been determined by diffraction and vibrational spectroscopic methods. Structural information on some of these materials is available only from infrared spectroscopic data. Detailed diffraction measurements are therefore needed to substantiate these observations and to determine the metric parameters.

A. Homopolar Compounds

The pseudohalide hydracids form a simple homopolar series which have been studied in some detail. The molecular parameters in the gas phase were computed from their vibrational-rotational spectra and are given in Table 1. The results show that the hydracids have a collinear X-Y-Z pseudohalide group but the H-X-Y angle varies from 114° for HN_3 to 180° for HCNO. In the condensed phase the hydracids are expected to exhibit a high degree of hydrogen bonding. The only data available is for HNCO at -125 °C, which exhibits an orthorhombic structure consisting of zig-zag N-H---N hydrogen bonds (7).

Table 1. *Structural parameters of pseudohalide hydracids, HXYZ*

Compound	H—XYZ angle	H—X Å	X—Y Å	Y—Z Å	Reference
HN_3	114.1°	1.021	1.237	1.133	(3)
HNCO	128.1°	0.987	1.207	1.171	(4)
HNCS	135.0°	1.000	1.216	1.561	(5)
HCNO	180.0°	1.027	1.161	1.207	(6)

B. Polar Crystals

The crystalline compounds of Group I A (including the ammonium ion) and Group II A elements are included in this category. The intra-ionic distances in the Group I A pseudohalides are expected to give the approximate intra-bond distances of the unpertubed pseudohalide ions, which are listed in Table 2 together with the values in partially ionic lattices. A molecular orbital calculation by *Kemmey et al* (8) on N_3^- in which the position of the N atoms are varied has yielded an equilibrium N-N separation of 1.20 Å which is in fair agreement with the experimental value in sodium azide. (*cf* Table 2).

Table 2. *Intra-ionic Distances of Pseudohalide Ions XYZ⁻ in Different Lattices*

Ion	Lattice	X—Y	Y—Z	Reference
N_3^-	NaN_3	1.178	1.178	(9)
N_3^-	BaN_6	1.168	1.164	(25)
		1.157	1.178	
N_3^-	αPbN_6	1.164	1.164	(33)
		1.193	1.160	
		1.177	1.166	
		1.213	1.147	
N_3^-	CuN_6	1.146	1.199	(35)
		1.213	1.098	
NCO⁻	KNCO	∼1.170	∼1.230	(11)
	AgNCO	1.195	1.180	(30)
CNO⁻	KCNO	∼1.190	∼1.400	(12)
	AgCNO (O)	1.090	1.250	(31)
	AgCNO (R)	1.120	1.200	(31)
NCS⁻	KNCS	1.150	1.690	(10)
	AgNCS	1.186	1.636	(32)

Table 3. *Structural Parameters of Polar Pseudohalides*

Compound	Class**		Phase and Temp (°C)	Space Group	Z*	Cell Consts (Å)					Ref.
						a	b	c	α	β	
LiN_3	M		25°	C_2/m–C_{2h}^2	2	5.63	3.32	4.98		107.4°	(9)
NaN_3	R	β	25°	R_{32} or R_3m–D_{3d}^5	1	5.49			38°43'		(9)
NaN_3	M	α	19°	C_2/m–C_{2h}^2	2	6.21	3.66	5.32		108°43'	(9)
$NaNCO$	R		25°	R_{32} or R_3m–D_{3d}^5	1	5.44			38°37'		(13)
$NaCNO$	R		25°	R_{32} or R_3m–D_{3d}^5	1	4.95			38°15'		(12)
$NaNCS$	O		25°	$Pnma$–D_{2h}^{16}	4	13.45	4.10	5.66			(14)
KN_3	T		25°	$14/mcm$–D_{4h}^{18}	4	6.09		7.056			(15)
RbN_3	T	I	25°	$14/mcm$–D_{4h}^{18}	4	6.36		7.41			(16)
RbN_3	C	II	315°	—	1	4.35		—			(17)
CsN_3	T	I	25°	$14/mcm$–D_{4h}^{18}	4	6.72	8.04				(18)
CsN_3	C	II	141°	—	1	4.53	—				(17)
$KNCO$	T		25°	$14/mcm$–D_{4h}^{18}	4	6.07		7.03			(15, 19)
$RbNCO$	T		25°	$14/mcm$–D_{4h}^{18}	4	6.35		7.38			(13)
$CsNCO$	T		25°	$14/mcm$–D_{4h}^{18}	4	6.71		8.04			(13)
$KCNO$	T		25°	$14/mcm$–D_{4h}^{18}	4	5.81		7.16			(12)
$KNCS$	O	I	25°	$Pcmb$–D_{2h}^{11}	4	6.63	6.66	7.58			(20)
$KNCS$	T	II	140°	$14/mcm$–D_{4h}^{18}	4	6.70		7.73			(21)
NH_4N_3	O		25°	$Pman$–D_{2h}^7	4	8.93	8.64	3.60			(22)
NH_4NCO	T		25°	—	1	3.64		5.57			(13)
NH_4NCS	M		25°	$P2_1/c$–D_{2h}^5	4	4.30	7.20	13.00			(23)
CaN_6	O		25°	$Fddd$–D_{2h}^{24}	8	11.62	10.92	5.66			(18)
SrN_6	O		25°	$Fddd$–D_{2h}^{24}	8	11.82	11.47	6.08			(24)
BaN_6	M		25°	$P2_1/m$–C_{2h}^2	2	9.59	4.39	5.42		99.75°	(25)

*Z = number of molecules in the unit cell
**M = monoclinic

R = rhombohedral, T = tetragonal, O = orthorhombic, C = cubic

The structural parameters of the ionic salts are given in Table 3. Comparisons can only be made in the case of the Group I A salts for which detailed structural information is available. The tetragonal Group I A azides, cyanates and fulminates are isostructural. In the case of the cyanates and fulminates the anions are randomly arrayed with respect to the end atoms but among the thiocyanates the anions are arranged alternately as NCS^- and SCN^-, probably as a result of the greater mass and lower electron affinity of the anion. The influence of the size of the cation relative to that of the anions is apparent among the alkali azides,

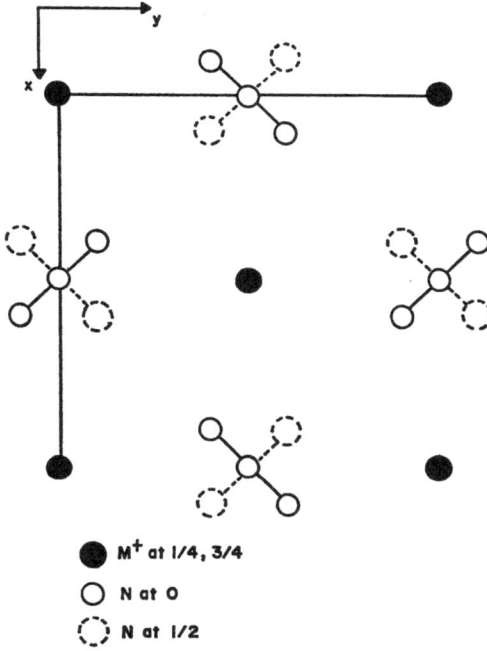

M^+ at 1/4, 3/4

N at 0

N at 1/2

Fig. 1. Projection of KN_3 type structure of space symmetry D_{4h}^{18} on the (001) plane

cyanates and fulminates. The pseudohalides of the larger alkali metal cations (K^+, Rb^+ and Cs^+) show eight fold co-ordination in a tetragonal distorted $CsCl$ type lattice (cf Fig. 1) and within a group the unit cell dimensions increase with increasing cationic radius. In the sodium salts, however, because of the smaller size of the cation, the co-ordination changes to 6:6, forming a more open rhombohedral lattice (cf Fig. 2a).

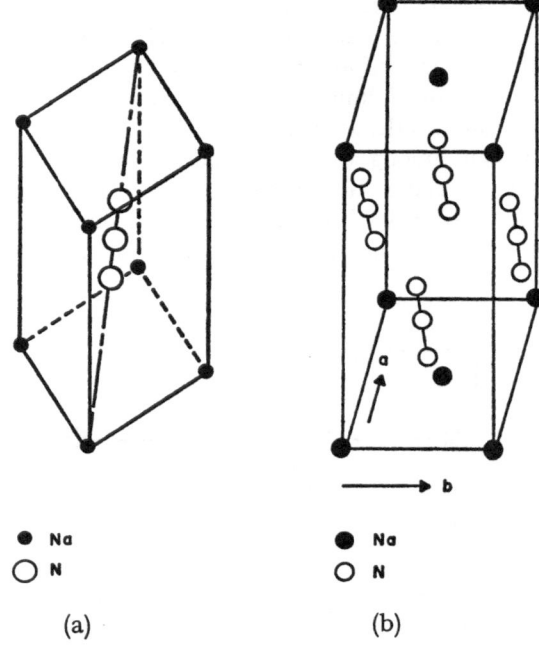

Na
N

(a)

Na
N

(b)

Fig. 2. a) Rhombohedral Unit cell of NaN$_3$ with Z = 1 and α = 38.7 °C b) Monoclinic Unit Cell of NaN$_3$ and LiN$_3$ with Z = 2, and β = 108.4 °

The structure of the ammonium salts show interesting inter-group variations. Ammonium azide has a lattice which is related to that of the tetragonal alkali salts but is distorted by hydrogen bonding to four nearest neighbor nitrogen atoms, to a structure of orthorhombic symmetry (22). Ammonium cyanate has a tetragonal structure with the cyanate ions lying along the tetrad axis because hydrogen bonding interactions are more specific towards the O atoms (13), while the thiocyanate salt has a structure of lower symmetry due to more extensive hydrogen bonding (23).

Among the Group II A salts, indications are that strontium azide (24) has an ionic lattice with the cations surrounded by eight near-neighbor azide ions. Barium azide, however, has a structure of lower symmetry (25) which shows some departures from a purely ionic lattice. This is evident from its non-totally reflecting reststrahlen infrared spectrum (26) and the presence of asymmetrical N$_3^-$ ions (cf Table 2).

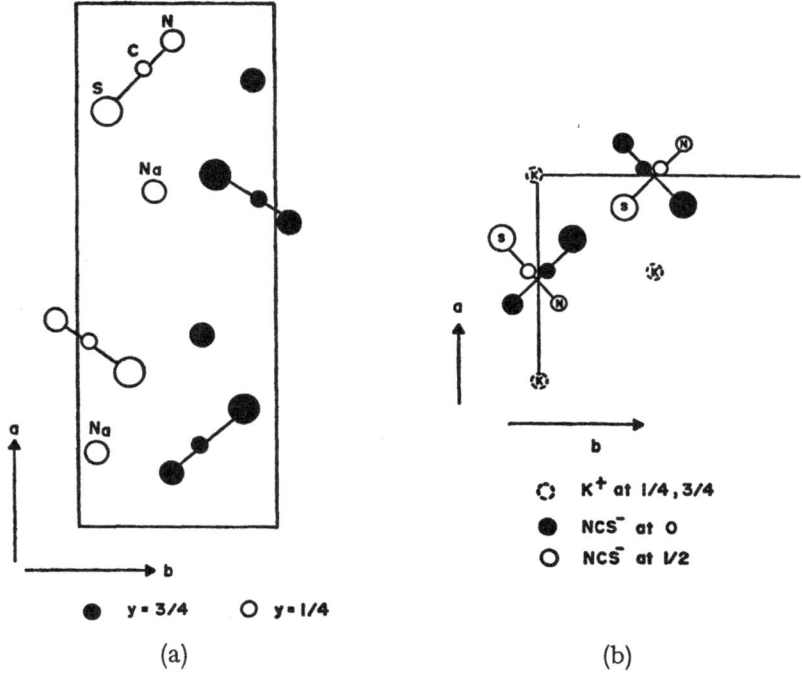

Fig. 3. a) Projection of orthorhombic NaNCS on the (001) plane b) Projection of orthorhombic KNCS on the (001) plane

C. Semi-Polar Crystals

The heavy metal pseudohalides fall in this group since covalent forces are partially present. The structural parameters listed in Table 4 show that thallous azide, fulminate and cyanate are isostructural with the tetragonal alkali metal salts, while thallous thiocyanate is isostructural with the room temperature phase of potassium thiocyanate. The lattice constants of the salts are all indicative of abnormally short metal-anion distances. The silver salts form an interesting series where the subtle interactions which result as a function of the electron affinity, electronic structure and size of the anions, stimulate somewhat predictable variations of crystal geometry. Silver azide is an example of a distortion of the tetragonal D_{4h} (Fig. 1) lattice due to covalent metal-anion interaction which lowers the space symmetry of the crystal to D_{2h} (29), while silver thiocyanate is mainly covalent with bidentate metal-anion chains

31

Table 4. *Structural Parameters of Semi-Polar Pseudohalides*

Compound	Class[b]	Phase and Temp (°C)		Space Group	Z[a]	Cell Consts (Å)					Ref.
						a	b	c	α	β	
TlN$_3$	T	I	25°	I4/mcm–D$_{4h}^{18}$	4	6.23		7.37			(16)
TlN$_3$	C	II	295°	—	—	—		—			(17)
TlNCO	T		25°	I4/mcm–D$_{4h}^{18}$	4	6.39		7.41			(13)
TlCNO	T		25°	I4/mcm–D$_{4h}^{18}$	4	6.23		7.32			(12)
TlNCS	O		25°	Pcmb–D$_{2h}^{17}$	4	6.80	6.78	7.52			(27)
CuN$_3$	T		25°	I4$_1$/a–C$_{4h}^{6}$	8	8.65		5.59			(28)
AgN$_3$	O		25°	Ibam–D$_{2h}^{26}$	4	5.59	5.91	6.01			(29)
AgNCO	M		25°	P2$_1$/m–C$_{2h}^{2}$	2	5.47	6.37	3.42		91°	(30)
AgCNO	O	I	25°	CmCm–D$_{2h}^{17}$	4	3.86	10.72	5.86			(31)
AgCNO	R	II	25°	R$\bar{3}$–C$_{3i}^{2}$	6	9.11			115°44'		(31)
AgNCS	M		25°	C2/c–C$_{2h}^{6}$	8	8.74	7.96	12.32		138°6'	(32)
PbN$_6$	O	α		Pcmn–D$_{2h}^{16}$	12	11.31	16.25	6.63			(33)
PbN$_6$	M	β		—	8	18.49	8.84	5.12		107°35'	(34)
CuN$_6$	O		25°	Pnma–D$_{2h}^{16}$	4	13.48	3.08	9.08			(35)
Hg(CNO)$_2$	O		25°	Pbca–D$_{2h}^{15}$	4	7.71	5.48	10.43			(36)

a) Number of molecules in the unit cell
b) T = tetragonal, O = orthorhombic, M = monoclinic, C = cubic, R = rhombic

(*cf* Fig. 4) (*32*). Silver cyanate (*30*) and fulminate (*31*) (*cf* Fig. 4) are likely to be more ionic than the corresponding azide and thiocyanate but have ordered structures indicative of some metal-anion bonding. The differences between the structure of silver cyanate and fulminate can be partially explained in terms of the nature of the metal-anion bond (*31*).

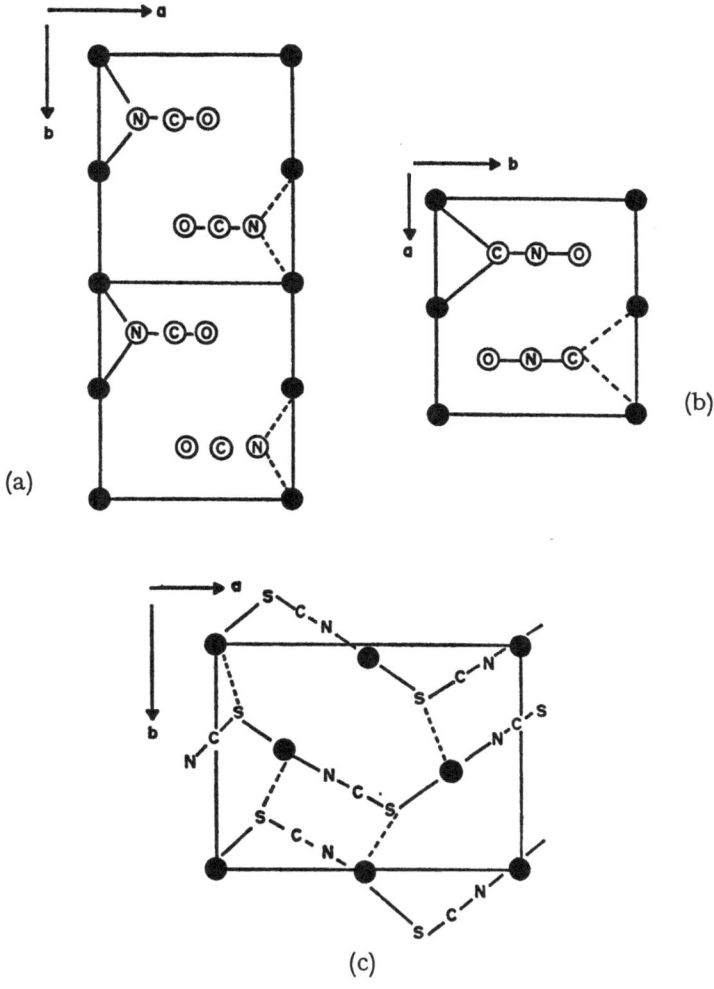

Fig. 4. Schematic projections of the unit cell of a) AgNCO b)AgCNO c) AgNCS down the *c* axis

Detailed crystal structures of only a few of the other heavy metal pseudohalides are available in the literature. Among them are, cuprous azide which has a relatively simple tetragonal lattice and cupric azide, mercuric fulminate and α-lead azide which have increasingly complex orthorhombic lattices. α-Lead azide has four types of anion sites of varying amounts of asymmetry (33) while cupric azide (35) and mercuric fulminate (12) have two such sites. The structure of cupric azide which is built up of distorted octahedra of asymmetric N_3^- ions about the central cupric ion is analogous to that of a transition metal complex.

D. Phase Transformations

Phase transformations have been observed in a number of azides in the temperature range $-40\ ^\circ C$ to $315\ ^\circ C$ and high pressure range 1 to 7 kbar. In all cases the high pressure phase belongs to a lattice of lower space group symmetry.

In sodium azide a transition occurs at $19\ ^\circ C$ (9) and also on application of ~ 1 kbar pressure (37a), in which the rhombohedral lattice transforms by a shearing motion of the azide ion layers to form a monoclinic unit cell (9). The latter is isostructural with the unit cell of lithium azide shown in Fig. 2b. Among the tetragonal rubidium, cesium and thallous azides a high temperature transformation in the range $151\ ^\circ C$ to $315\ ^\circ C$ to a cubic structure takes place (17), while at $-40\ ^\circ C$ a transition to an orthorhombic structure has been recently established for thallous azide (38). In the range 4 to 6 kbar, *Pistorius* (39) has observed pressure induced polymorphs of rubidium, cesium and thallous azides which are expected to be isostructural with the low temperature phase in thallous azide.

Mueller and *Joebstl* (17) have argued that enhanced thermal motions at higher temperatures lead to a statistical disordering of the N_3^- ions in the cubic phase. The increase in entropy calculated from the transformation enthalpy is of the order of 1.8 cal/mol. deg. for cesium and rubidium azide, which is close to the value of R $ln\ 3 = 2.13$. This would suggest that the anions are oriented at random parallel to the three edges of the cubic unit cell (17). The increase in entropy for the transition in thallous azide is however 1.0 cal/deg. mol., which can be explained by a lowering of the disordering of the anions due to the comparatively large polarizibility of Tl^+ (17). Among the alkali metal azides, the tetragonal to cubic transition temperature decreases linearly with increasing size of the cation. The transition point is therefore expected to be rather high for potassium azide and has not been observed because the melting point is below this temperature.

The change of entropy in the high pressure transition is close to 1.38 cal/mol. deg. (R $ln\ 2$) for rubidium and cesium azides, while it is close to 2.18 cal/mol. deg. (R ln 3) for thallous azide (*39*). The entropy change of 1.38 cal/mol. deg. can be accounted for in terms of crystallographically non-equivalent anions, which suggests a lattice isostructural with that of silver azide (*39*). Evidence for this has been obtained for thallous azide from its lattice mode infrared spectrum below the transition temperature (*40*). Interestingly enough the medium infrared spectrum of a single crystal of thallous azide shows the appearance of the forbidden symmetric stretching mode of N_3^- at 1325 cm^{-1} below the transition temperature which neatly suggests the existence of non-centro-symmetric N_3^- ions (*40*). The larger entropy change for the thallous azide transition can be attributed to a contribution from the vibrational entropy term, which is expected to be more sensitive to geometrical changes for lattices containing highly polarizable ions (*39*).

Among the thiocyanates another order-disorder phase transition is possible. This involves a transition from an ordered structure similar to potassium thiocyanate to a disordered structure isostructural with potassium cyanate. This has been observed in potassium thiocyanate at 145 °C but has not been observed up to 300 °C in the nearly isostructural sodium salt. *Iqbal* (*41*) has suggested an explanation for this behaviour in terms of the greater anisotropy of the inter-ionic interaction potential in potassium thiocyanate compared with the sodium salt.

A great variety of complex phase transformations are possible in the heavy metal pseudohalides. The simplest observed is a sluggish high temperature transformation at 185 °C in silver azide which is probably analogous to the tetragonal-cubic transition in the alkali metal salts (*42*). A second polymorph of silver fulminate (*31*) and β and γ forms of lead azide (*2*) have been isolated. These form close to room temperature and their growth depends on crystallizing conditions. It is interesting to note that the rhombohedral polymorph of silver fulminate has a grossly different geometry compared with the normal orthorhombic form of the salt. The detailed crystal structures of the β and γ forms of lead azide are not known at present.

III. Vibrational Energy Levels and Bonding Forces

The vibrational energy levels provide insights into intra- and inter-molecular bonding forces and thermal properties while related parameters like the lattice energy, heat of formation and dielectric constants determine the magnitude of the binding forces and the degree of lattice ionicity. These properties will be considered in terms of the degree and nature of the bonding forces in the inorganic pseudohalides.

Table 5. *Vibrational Frequencies and Force Constants of Pseudohalide Ions*

Ion	Solvent	$v_1{}^{a)}$ (Σ_g^+ or Σ^+)	v_2 (π_u or π)	v_3 (Σ_u^+ or Σ^+)	Force Constants dynes cm^{-1} $\times$ 10^5			Ref.
					K_{X-Y}	K_{Y-Z}	$K_{\delta/l_1 l_2}{}^{b)}$	
N_3^-	KCl		641.92	2048.93				(46)
	KN$_3$	1340.00	640.00	2068.40	12.520	12.520	0.5721	(45)
		1352.00	645.40					
NCO$^-$	KBr	1254.03^c	629.64^c	2182.60^c	15.879	11.003	0.5086	(43)
NCS$^-$	KNCS	763.60^c	484.40^c	2087.70^c	15.950	5.180	0.300	(44)
			494.10					
CNO$^-$	KCNO	1105.00	469.00	2082.00	13.140	7.800	0.2930	(12)
			473.80					

a) v_1, v_2 and v_3 are the symmetric stretching, bending and antisymmetric stretching modes of XYZ$^-$
b) l_1 and l_2 are the intraionic distances of X–Y and Y–Z
c) These frequencies have been corrected for anharmonicity

A. Molecular Vibrations

In the isolated pseudohalide ions the normal molecular vibrations are distributed as one mode each of symmetric (ν_1) and anti-symmetric (ν_3) stretching motions and a doubly degenerate pair of bending (ν_2) motions. In the centrosymmetric N_3^- ion, ν_1 is active only in the Raman spectrum. The infrared and Raman spectra of N_3^-, NCO^- and CNO^- cannot be reliably determined because of the ease with which these ions hydrolyze in solution. Values fairly close to the free ion frequencies can, however, be obtained from the spectrum of the corresponding alkali metal salt or of the ion substituted into an alkali halide lattice. These frequencies are listed in Table 5, together with the stretching and bending force constants calculated using a simple valence force field. The bending force constants are comparatively low for CNO^- and NCS^-, which suggests that the O and S atoms respectively are rather weakly bonded in the two anions.

Six vibrational modes are expected for the pseudohalide hydracids for which the normal mode description and energies together with the molecular force constants are shown in Table 6. It is worth noting here that the N-H bond stretching force constant for HN_3 is lower than that for HNCO and HNCS. This indicates that the latter are thermodynamically more stable compared with HN_3.

Table 6. *Vibrational Frequencies and Force Constants of HXYZ Hydracids (cm⁻¹)*

			HN_3 (47)	HNCO (48)	HNCS (48)	HCNO (49)
(ν_1)	H—X	Stretch	3336	3536	3536	3335
(ν_2)	X—Y—Z	Antisymmetric Stretch	2140	2274	1963	2190
(ν_3)	X—Y—Z	Symmetric Stretch	1274	1327	860	1251
(ν_4)	H—X—Y	Bend	1150	797	817	
(ν_5)	X—Y—Z	Bend	522	572	469	
(ν_6)	X—Y—Z	Bend	637			538
	K_{X-Y} DYNES $\times 10^5 \times cm^{-1}$		10.1a)	14.0	13.2	
	K_{Y-Z} DYNES $\times 10^5 \times cm^{-1}$		17.3	15.0	7.27	
	K_{H-X} DYNES $\times 10^5 \times cm^{-1}$		6.2	6.9	7.0	

a) Force constant values after *Orville-Thomas (50)*. Detailed calculations on HN_3 have been performed by *Thompson* and *Fletcher (51)*.

B. Vibrations in Crystals

The vibrational potential energy of a crystal consisting of a cation and molecular anion is made up of the static and dynamic terms due to the internal modes of the anion together with the terms due to the lattice modes. The static term accounts for perturbations of the internal mode frequencies and breakdown of degeneracies if the site group is of a lower symmetry than the molecular point group, while the dynamical terms most importantly induce correlation field splitting associated with the site to factor group transformation. In the absence of the correlation field, the perturbing potential (V') for an isotropic lattice is given by the relation (46):

$$V' = V_i + V_c + V_s \tag{1}$$

where V_i, V_c and V_s respectively represent the inductive, coloumbic and repulsive contributions to the perturbing potential. When NCO^- and N_3^- are introduced into cubic alkali halide lattices, small perturbations due to a potential of type V' occur. Interestingly enough, the internal modes also decrease in frequencies with decreasi ng ionicity of the solvent lattice (43, 46). Similarly in the anisotropic tetragonal lattice a decrease of 100 cm^{-1} in ν_3 of N_3^- has been observed as one goes from the largely ionic lattice of potassium azide to the partially ionic structure of thallous azide (45, 52). In the more complex crystal structures of barium azide and mercuric fulminate, ν_3 is larger than in the corresponding alkali metal salt (26, 12). This is probably associated with the partial formation of σ-type metal to anion bonds in these salts.

The lattice modes or phonons due to the center of mass librations of the anions and translations of both cations and anions, occur at the low frequency end of the vibrational spectrum. The translational modes in particular, unlike the internal modes show considerable dispersion in energy-momentum space. The in-phase or acoustic vibrations tend to zero energy at the zone center ($\vec{k} = 0$)[1]), but disperse to higher energies near the zone boundary. The acoustic phonons in a crystal of potassium azide have been determined at different points in the Brillouin zone by means of coherent neutron inelastic scattering (53). The experimental curve and the values calculated using a rigid ion model are shown in Fig. 5 (53).

The symmetry allowed modes at $\vec{k} \sim 0$ have been determined using infrared and Raman spectroscopic techniques by *Iqbal et al* for a number of alkali pseudohalides (19, 41, 54, 55), and ammonium (56), barium (26), α-lead (57) and thallous azide (45). The $\vec{k} \sim 0$ dependence of the first

---

[1]) k ($= 2\pi$/wavelength) is the phonon wave vector.

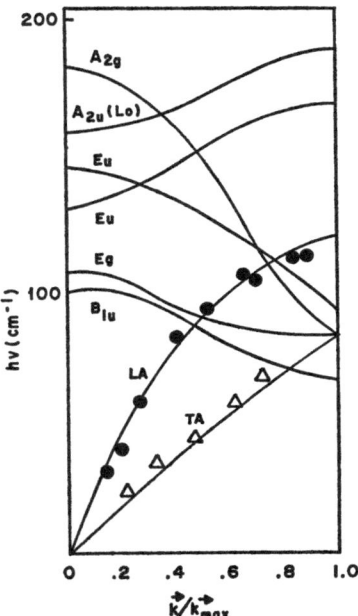

Fig. 5. The dispersion curves for the translational optic and acoustic modes of potassium azide in the (001) direction. TA, LA and LO indicate the transverse acoustic, longitudinal acoustic and longitudina optic modes respectively and Δ and 0 represent experimental points and the full lines the calculated curves

order bands break down for multiphonon spectra and it has been possible to make some explicit assignments of modes at points of high phonon densities from (internal and external) combination bands in the infrared spectra (*19, 41*). A statistical distribution of modes throughout the Brillouin zone analogous to the multiphonon spectra is observed in incoherent neutron inelastic scattering studies. Data along these lines have been obtained by *Trevino et al (58)* for sodium and ammonium azide. Due to the large hydrogen scattering cross-section of low energy neutrons, it is possible to detect the torsional mode frequency of the ammonium ion in ammonium azide by this method. The mode is found to be fairly anharmonic with the first and second excited energy levels at 400 and 700 cm^{-1} respectively above the ground state. The translational motions of the ammonium ion in ammonium azide, however, occur at approximately one-half the energy of the torsional mode (*56*).

The $\vec{k} \sim 0$ infrared and Raman active optic translatory and libratory modes of the pseudohalides which belong to space group D_{4h}^{18} can be treated in a comparative fashion since experimental data for a whole series is presently available. The data is shown in Table 7 (*59a*). The factor group normal modes can be drawn up from group theoretical considerations and are depicted in Fig. 7. A rigid ion model has been used to calculate the dispersion of these modes in potassium azide. The calculated curves are shown in Fig. 5 (*53*).

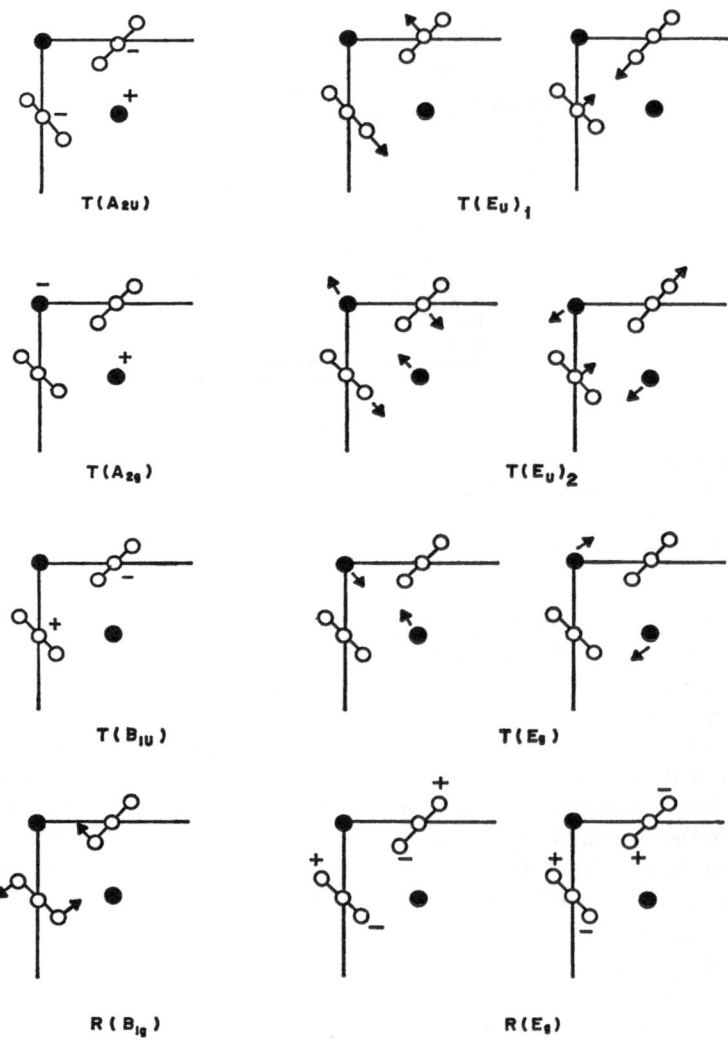

Fig. 6. Factor group optic translational (T) and librational (R) lattice modes for crystals belonging to space group D_{4h}^{18}. The letters within brackets indicate the symmetry species for each vibration

Table 7. *Raman and Infrared active* $\vec{k} \sim 0$ *Phonons in Tetragonal Pseudohalides of* D_{4h}^{18} *Symmetry*

Compound	Translational Modes (cm^{-1})[a]			Librational Modes[a] (cm^{-1})		Ref.
	E_u	A_{2u}	E_g	B_{1g}	E_g	
KN$_3$	166, 130	80 [b]	102	150	147	(54,55)
RbN$_3$	144, 96	76	65	140	126	(54,55)
CsN$_3$	124, 90	46	42	153	108	(60)
TlN$_3$	128, 96	83 [b]	36	173	51	(45)
KNCS	—	—	58	117	80	(61)
KNCO	170, 146	74	109	125	123	(19)

[a]) The modes have been reassigned in some cases consistent with recent lattice dynamics work.
[b]) These bands might also arise due to multiphonon modes.

A comparison of the $\vec{k} \sim 0$ phonon energies for these crystals indicate that motions within the plane of the anion (symmetry B_{1g} and E_u) involve more energy than motions perpendicular to this plane (symmetry E_g and A_{2u}). The data is therefore indicative of weaker interactions parallel to the c-axis. Further confirmation of this has come from compressibility measurements on the metal azides (59 b). The increasing separation of the librational E_g from the librational B_{1g} mode also suggests that the former is mixed to a large degree with the translational E_g mode. A correlation between the librational E_g to B_{1g} mode separation and lattice ionicity can be made since it is found that the separation is the least for potassium cyanate (ionic) and the largest for thallous azide (semi-ionic) (45). It is also worth noting that the libratory modes in potassium cyanate and thiocyanate are 25 and 34 cm^{-1} respectively lower than the corresponding modes in potassium azide. This is partially a reflection of the lowered reduced mass since librations essentially involve the end atoms of the anions.

C. Thermal Properties

The heat capacity (C_v) and heat conductivity (λ) of crystals depend respectively on the vibrational density of states weighted by a *Boltzmann's* distribution factor and the anharmonic terms in the vibrational potential energy. C_v has been found to be in the range 0.100 to 0.117 cal./g./°C between 100—250 °C for crystals as widely varying in lattice geometry as mercuric fulminate, silver azide and lead azide (62). This is

41

probably not surprising since a large number of the vibrational modes involve excursions of anions of the same reduced mass in inter-ionic potential fields of approximately the same order of magnitude. One would therefore expect the elastic wave velocity (V_0) in a whole series of ionic and semi-ionic pseudohalides to be nearly of the same order of magnitude.

The thermal conductivity (λ) is given by the relation (63):

$$\lambda = (^1/_3) V_0 C_v \Lambda \qquad (2)$$

where Λ is the phonon mean free path. The relative thermal conductivities of a series of materials like the metal pseudohalides would therefore be essentially dependent on Λ. In a pure and infinite crystal Λ tends to infinity if a harmonic vibrational potential is assumed. However, in real crystals Λ is finite because of impurities, geometrical factors and sizeable cubic and higher order (anharmonic) terms in the vibrational potential energy. The resistance to phonon flow can then be visualized as a three body scattering process of the Umklapp type which is inversely proportional to temperature, above the characteristic Debye temperature of the solid.

The average thermal conductivities of some of the metal azides and mercuric fulminate have been measured in the temperature range 45—80 °C and the results together with the normalized values at 25 °C are given in Table 8 (62). It is seen that the thermal conductivity decreases with increasing covalency and decreasing lattice symmetry and in the case of the azides, the values decrease with an increase in the molecular weight of the salts. Since the binding forces and the lattice constants change, the decrease in thermal conductivity cannot be solely attributed

Table 8. *Thermal Conductivity (λ) of Pressed Pellets of Metal Pseudohalides*[a])

Compound	Mean Temp °C	Density of Pellet gm/cc	Thermal Conductivity c.g.s. units	Values Normalized to 25 °C c.g.s.
NaN$_3$	68	1.765	25×10^{-4}	$\sim 68 \times 10^{-4}$
BaN$_6$	80	2.75	11×10^{-4}	$\sim 35 \times 10^{-4}$
TlN$_3$	65	4.60	8×10^{-4}	$\sim 21 \times 10^{-4}$
α-PbN$_6$	45	3.62	4×10^{-4}	$\sim 7.2 \times 10^{-4}$
Hg(CNO)$_2$	—	—	2.9×10^{-4}	—

[a]) Values taken from *Bowden* and *Yoffe* (62)

to an increase in atomic mass as one goes from sodium to lead azide. A correlation between the anisotropy of the binding forces and Λ is therefore implied.

D. Lattice Energy and Heat of Formation

The lattice energy determines the magnitude of the binding forces in ionic and partially ionic crystals and may be defined as the increase in internal energy at absolute zero accompanying the separation of the constituent ions to positions where they are infinitely removed from one another. The lattice energy (U) can be represented by the expression:

$$U = \frac{1}{2} \sum_{l\,k,\,l'\,k'} \left[\frac{Z_k\,Z_{k'}}{r} + b \, \exp. \left(\frac{-r}{\varrho} \right) + U_L \right] \tag{3}$$

The first term represents the coulombic interaction between the ions with charge Z_k and $-Z_{k'}$ at an equilibrium lattice parameter "r" from each other; the second term is the repulsive term in which 'b' and 'ϱ' are constants and the third term represents the contribution from London forces. The latter can be written in terms of the polarizibilities α_1 and α_2 of two ions 'r' apart as:

$$U_L = \frac{3\,\alpha_1\,\alpha_2}{2r^6} \, \frac{\varepsilon_1\,\varepsilon_2}{\varepsilon_1 + \varepsilon_2} \tag{4}$$

where ε_1 and ε_2 are energies characteristic of the oscillators in the two ions. *Gray* and *Waddington (64)* have employed a term-by-term calculation using the above relations to calculate the lattice energies of the tetragonal alkali azides. These values are given in Table 9 and show a decrease with increasing lattice constants in agreement with Eq. (3). The lattice energies of a series of metal azides were also determined by *Gray* and *Waddington (64)* from measured thermochemical quantities. The values for the divalent salts are larger because of the increased charge while the lattice energies of the monovalent heavy metal azides increase because of the larger cationic polarizibilities. It is however interesting to note that the melting points of thallous and silver azides are lower than that of potassium azide. This suggests a diminished stability of the crystals of the heavy metal salts relative to the melt. Recently *Gora (65)* has calculated the Madelung energy in the more complex barium and α-lead azide crystals. On assuming a symmetrical anionic charge distribution the calculated values were close to the experimental values of *Gray* and *Waddington (64)*. This suggests that the anion charge clouds in these crystals are not largerly perturbed. This is in disagreement with structural results.

Table 9. *Lattice Energies (kcal/mole) at 25 °C of Metal Pseudohalides. Determined by Different Methods*

Compound	Calculated Values		From Hydration Heats			Ref.
	Madelung Term	Lattice Energy	Ionic Radius	Lyotropic Number	From Heat of formation	
LiN_3	—	—	191.0	194.0	—	(67)
NaN_3	199.0	—	173.0	175.0	—	(67)
KN_3	161.7	157.0	155.0	157.0	—	(67)
RbN_3	149.9	150.0	149.5	151.5	—	(67)
CsN_3	141.8	143.0	143.5	145.5	—	(67)
NH_4N_3	—	—	—	—	175.0	(64)
CuN_3	—	—	—	—	227.0	(64)
AgN_3	—	—	—	—	204.0	(64)
TlN_3	—	—	—	—	163.3	(64)
CaN_6	—	—	—	—	517.4	(64)
SrN_6	—	—	—	—	493.8	(64)
BaN_6	510.0	—	—	—	468.8	(64, 65)
α-PbN_6	538.0	—	—	—	516.4	(64, 65)
NaNCO	—	—	176.0	—	—	(67)
KNCO	~163.0	158.0	156.0	—	—	(67)
NaNCS	—	—	161.0	—	—	(67)
KNCS	—	—	143.0	—	—	(67)

The heat of formation measures the work required to dissociate a crystal into an assembly of neutral atoms and one arrives at this quantity by means of the Born-Haber cycle. This is given by the following Eq.:

$$\Delta H_f^0 = [\Delta H_f^0 \text{ (radical)} - (E)] + [(I + L) - (U + 2\ RT)] \tag{5}$$

where ΔH_f^0 (radical) and E are the heat of formation and electron affinity of the pseudohalide radical, I and L are the ionization potential and the sublimation heat of the metal and U is the lattice energy. From the above Eq. it can be seen that for the metal salts, intra-group differences in ΔH_f^0 would depend on the second term whereas inter-group differences would depend essentially on the first term. This is reflected in the experimental values for a whole range of metal pseudohalides shown in Table 10. The values indicate that the exothermicity of decomposition in these materials would increase in the order cyanate < thiocyanate < fulminate < azide. The fulminates are however the most sensitive to thermal shock and this is probably due to exothermic side reactions (see Section V).

Table 10. *Heats of Formation of Crystalline Pseudohalides (2, 27)*

Salt	ΔH_f^0 (kcal/mole)
NaN_3	5.1
KN_3	—0.33
KNCO	—98.5
KNCS	—48.02
NH_4N_3	26.79
NH_4NCO	—74.7
NH_4NCS	—20.0
TlN_3	55.8
AgN_3	74.2
AgNCO	—21.2
AgNCS	21.0
AgCNO	43.2
CuN_3	67.23
CuCNO	26.3
$Hg(N_3)_2$	141.5
$Hg(NCS)_2$	48.0
$Hg(CNO)_2$	64.5
α-$Pb(N_3)_2$	115.0
CaN_6	11.0
SrN_6	1.7
BaN_6	—5.3

E. Dielectric Behaviour

In ionic and partially ionic crystals optic vibrations are associated with strong electric moments and hence can interact directly with the transverse electric field of incident infrared electromagnetic radiation. In terms of the phenomenological theory of infrared dispersion, if $\vec{E}$, $\vec{P}$ and $\vec{D}$ are the electric field, polarization and displacement vectors respectively, then

$$\vec{E} + 4\pi\vec{P} = \vec{D} = \varepsilon\vec{E} \qquad (6)$$

where ε is the dielectric constant of the material. ε can be expressed in terms of the static (ε_0) and high frequency (ε_∞) dielectric constants by the relations:

$$\varepsilon = \varepsilon_\infty + \frac{(\varepsilon_0 - \varepsilon_\infty)\,\nu_0^2}{\nu_0^2 - \nu^2}; \quad \varepsilon_\infty = n^2 \qquad (7)$$

where ν_0 is the dispersion frequency and n is the refractive index for $\nu > \nu_0$. ε_0 is greater than ε_∞ for ionic materials because of the extra contribution from ionic displacements due to the external field.

The static dielectric constant (ε_0) of the alkali azides (*cf.* Table 11) is of the order of 6.5 while ε_∞ is 2.3. These values are of the same order as those of the alkali halides. Both the high and low frequency dielectric constants increase in the case of thallous, silver and cuprous azides. This is likely to be due to the increasing polarizibility of the cations and a reflection of the decreasing ionicity of the lattices. The ε_0 value for thallous azide and silver fulminate are however surprisingly high compared with the other azides.

Table 11. *Dielectric Constants of Ionic and Semi-Ionic Pseudohalides*

Compound	$\varepsilon_\infty (= n^2)$	ε_0	Ref.
NaN_3	2.2	6.4	(2)
KN_3	2.3	6.9	(2)
TlN_3	3.3	11.5	(2)
AgN_3	4.9	9.4	(2)
CuN_3	3.3	9.3	(2)
$Ba(N_3)_2$	—	7.7	(2)
$Sr(N_3)_2$	—	8.3	(2)
$AgCNO$	6.8	11.8	(12)
$Hg(CNO)_2$	2.4	—	(12)
$AgNCO$	$\sim$4.0	—	(62)

As a consequence of infrared dispersion by ionic crystals, electromagnetic radiation with frequencies in the vicinity of the dispersion frequency undergoes selective reflection or reststrahlen. For ideally ionic crystals and normally incident radiation, reflection is ca. 100% in terms of the damped oscillator model (37b). Investigations by *Mitra* (66) on the alkali azides show a decreasing percentage reflectance in the region of the lattice modes with increasing lattice constants of the salts. Hence the degree of ionicity of these materials decreases as one goes from sodium to cesium azide. In the case of barium azide reflectivity measurements show that the salt is less ionic than the alkali azides (26).

IV. Electron Energy Levels

The problem of describing how atoms are held together in molecules can be best tackled in terms of electronic wave functions. One starts with wave functions which are localized on individual atoms in the molecule and then proceeds to combine these functions in various trial combinations. These combinations are called molecular orbitals and are used to describe molecular electron energy levels. The ground states of the pseudohalide ions and their hydracids can be readily treated in terms of this theory. Problems arise in the interpretation of the vacuum ultra-violet electronic spectra of the former since the measurements can be made only in the solid state. However, to a first approximation one can neglect the crystal field perturbation of the molecular levels in the essentially ionic alkali metal salts and the discussion below treats observed spectra as arising solely from the molecular ion electron energy levels. In the case of the crystal spectra of the more complex divalent and heavy metal salts such an approximation is not valid and one has to take into account both the crystal field and the solid state band nature of the electron energy levels. Here the definitive interpretation of experimental spectra needs to be coupled with band structure calculations.

A. Electronic Levels in Pseudohalide Ions and Hydracids

The pseudohalide ions have sixteen valence electrons. *Walsh (68)* has shown that the most stable configurations for such molecules is linear. The lowest energy orbitals for an XYZ^- entity can be described as follows:

(i) Two lone pair 's' orbitals on the end X and Z atoms

(ii) Two σ type 'sp' hybridized bonding orbitals localized in the X—Y and Y—Z bond distance.

(iii) A π-bonding orbital built by the in-phase overlap of a $p\pi$ atomic orbital on each of the three atoms

(iv) Another π orbital built by the out-of-phase overlap of a $p\pi$ atomic orbital on the X and Z atoms. It has zero amplitude at the central atom and is localized on the end atoms. It is weakly X → Y antibonding and X → Y and Y → Z non-bonding.

(v) A π^*-orbital built by the in-phase overlap of a $p\pi$ atomic orbital on each of the X and Z orbitals and overlapping out-of-phase with a $p\pi$ orbital on the Y atom. It is X → Z bonding and X → Y antibonding.

The formation of the pseudohalide hydracids results in a perturbation of the XYZ^- molecular orbitals due to protonation. In HN_3, HNCO and HNCS the perturbation leads to a 'sp^2' hybridization of the lone pair on the N atoms. This overlaps with the 's' type atomic orbital in H to form the hydracids. The H—XYZ bond angles are therefore expected to be closer to 120° (*cf.* Table 1). The π-orbital degeneracies breakdown because of the lowering of the molecular symmetry from $C_{\infty v}$ and $D_{\infty h}$ to C_s and the orbital energies decrease somewhat in accordance with the greater stability of the hydracids compared with the corresponding ions. In the case of HCNO the lone pair on C is 'sp' hybridized and hence the H—XYZ angle is close to 180° (*cf*. Table 1).

Several calculations of the ground state orbital energies of N_3^- (*73, 74, 69*), NCO⁻, NCS⁻, HN₃ and HNCO (*69*) have been made. *McGlynn et al.* (*69*) have calculated the energies for all five molecules using a Mulliken-Wolfsberg-Helmholz computation. Their results are shown in Fig. 7. The energies in NCO⁻ and NCS⁻ are fairly close and hence the

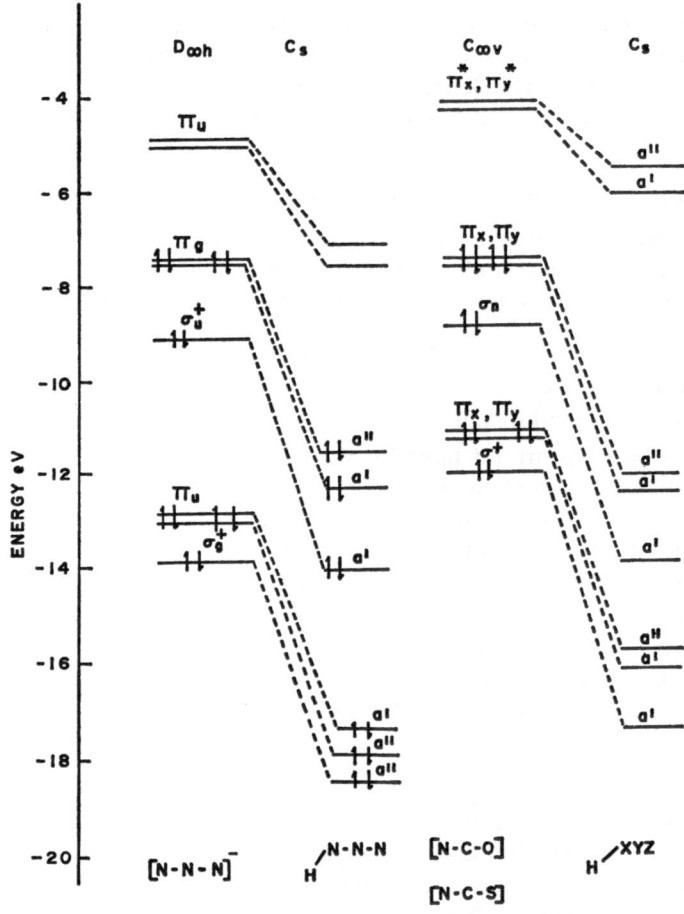

Fig. 7. The ground state molecular orbitals in N_3^-, HN₃, NCO⁻ and NCS⁻. These are drawn schematically (for exact values see ref. *69—72*) with respect to the calculated orbital energy values of *McGlynn et al* (*69*) (also see text). The scheme marked HXYZ represent the expected ordering for HNCO and HNCS

average values are shown. The π-bond characters of the pseudohalide ions and their hydracids have been calculated by *Wagner (11)*. His results are shown in Table 12 and 13. These indicate a larger inter-bond π-charge distribution asymmetry in CNO⁻ and NCS⁻. Among the hydracids, however, the π-charge distribution of the pseudohalide groups are symmetrical only for HNCO and HNCS. The central atoms are seats of positive charge in the pseudohalide ions. This is nicely illustrated for N_3^- by a well resolved doublet corresponding to the N 1s electron binding energies in the photoelectron spectrum of crystalline sodium azide *(75)*.

Table 12. *Bond Characters of Pseudohalide Ions (After Wagner (11)).*

Ion XYZ⁻	π-Electron Atomic Charge			π-Bond Order	
	Q_x	Q_y	Q_z	P_{XY}	P_{YZ}
N_3^-	—0.8060	+0.6121	—0.8060	1.3874	1.3874
NCO⁻	—0.6792	+0.0396	—0.3604	1.6413	1.1422
CNO⁻	—1.1663	+0.6516	—0.4852	1.7402	0.9221
NCS⁻	—0.4826	+0.1934	—0.7108	1.8243	0.7964

Table 13. *Bond Characters of the Pseudohalide Hydracids (After Wagner (11))*

Acid H XYZ	π-Electron Atomic Charge			π-Bond Orders		Dipole Moment Along XYZ axis Debye Units
	Q_x	Q_y	Q_z	P_{XY}	P_{YZ}	
HN_3	—0.7332	+0.8991	—0.1659	0.6121	1.5429	1.1[a]
HNCO	+0.2908	—0.0372	—0.2536	1.5917	1.2104	1.59[b]
HNCS	—0.1575	+0.6709	—0.5134	1.5001	1.1400	1.72[b]
HCNO	—0.1783	+0.6412	—0.4630	1.7291	0.9388	3.06[c]

[a] *J. Alster* (private communication)
[b] *W. J. O. Thomas*, "Structure of Small Molecules", p. 134, Elsevier (1966)
[c] Ref. (6)

The observed electronic spectra of N_3^- *(72, 76, 79)*, NCO⁻ *(71, 77)*, NCS⁻ *(70)* and CNO⁻ *(12, 78)* and HN_3 *(72)* can be interpreted in terms of the excited states of these molecular entities. Anionic excited states of types $^{1,3}\Sigma^+$, $^{1,3}\Delta$ and $^{1,3}\Sigma^-$ result due $\pi_g \rightarrow \pi_u$ or $\pi_{x,y} \rightarrow \pi^*_{x,y}$ transitions, while $^{1,3}\Pi$ can result due to $\sigma_U^+ \to \pi_u$ or $\sigma_N \to \pi_{x,y}$ transitions The ordering of the excited states in the anions and HN$_3$ together with the observed transitions are shown in Fig. 8, based on the work of *McGlynn et al.* While it appears that this ordering of states is probably correct for the non-centrosymmetric ions, in N$_3^-$ by analogy with CO$_2$, the $^1\Delta_u$ state might be lower than the $^1\Sigma_u^-$ state. Indications are also that all the excited states, except the $^1\Sigma_u^+$ and $^1\Sigma^+$ states, are bent to varying degrees (*69—72, 74*). The ordering of states suggested by *McGlynn et al* can, however, be nicely correlated with the observed spectrum of HN$_3$. Observations made by *Iqbal* and *Yoffe (12)*, *Deb (78)* and *Iqbal (77)* on CNO$^-$ and NCO$^-$ can also be fitted into *McGlynn's* scheme as shown in the same Fig. thus giving a satisfying correlative picture of the excited state transitions in the anions.

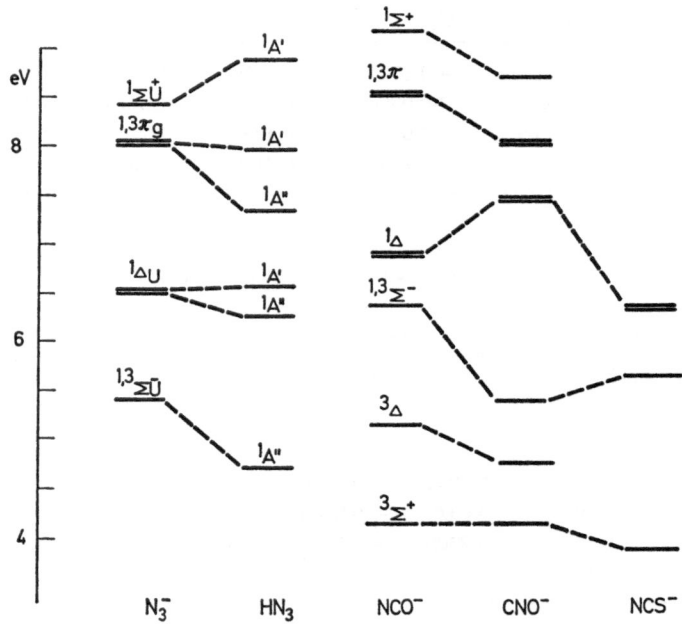

Fig. 8. Observed electronic transitions in N$_3^-$, HN$_3$, NCO$^-$, NCS$^-$ and CNO$^-$ plotted in accordance with the ordering of excited states suggested by *McGlynn et al (69—72)*

B. Electronic Levels in Crystals

In the crystalline state the molecular levels of pseudohalide ions and the atomic levels of the cations overlap to form bands of filled electronic energy levels, with an energy gap between the top of the filled levels (valence band) and the essentially unfilled higher energy conduction levels. The electronic wave functions, therefore, depend on the wave vector $\vec{k}$ and a lattice periodicity function. Optical transitions then occur from the valence band to excited levels in the energy gap.

In the alkali metal pseudohalides the contribution of cationic wave functions to the valence band structure can be neglected. The optical absorption spectra can therefore be correlated to transitions involving excited states of the anions. However, one can see solid state effects like the superposition of vibronic structure on the molecular symmetry forbidden transition at 5.39 eV in the crystal spectra of the alkali metal azides (76). In the more complex heavy metal and divalent azides, a whole range of optical transitions can occur both due to crystal field effects and the enhanced contributions from cationic states to the valence band. Detailed spectral measurements on α-PbN_6 (80), TlN_3 (81), AgN_3 (82), $Hg(CNO)_2$ (12) and $AgCNO$ (12) have been made but the level assignments can at best be described as tentative since band structure calculations on these materials are not available at present.

V. Relationship between Structure and Stability

All the crystalline metal fulminates and the heavy metal azides explode and detonate on the application of thermal shock and in some cases in the presence of an electric field, while the heavy metal cyanates and thiocyanates only deflagrate at high temperatures. According to *Bowden* and *Williams* (83) the stability to thermal shock increases in the order fulminate <azide <thiocyanate <cyanate and within a group (except in the case of the fulminates) the heavy metal salts are more thermally unstable. During the last decade *Yoffe* and co-workers have tried to explain the thermal and photochemical behaviour of these meterials in terms of their electronic structure. This work has been extensively reviewed by *Yoffe* (2) and will not be considered here. We will however explore briefly a few ideas which qualitatively correlate the geometry of the electronic and atomic configurations with the kinetic stability of these materials.

The geometry of the electronic configuration in the pseudohalide ions can be treated in terms of Linnett's non-paired spin orbital approach (84). Six possible resonance structures can be drawn up as shown below (84). The lines indicate a pair of electrons of opposite spin in the same spatial orbital, the dotted lines indicate a pair of different spatial orbitals while o and x represent electrons of different spin.

(I) $X = Y = Z$

N_3^-; $-1, +1, -1$

NCO^-; $-1,\ 0,\ 0$

CNO^-; $-2, +1,\ 0$

(II) $---X-Y\equiv\equiv\equiv Z-$

N_3^-; $-2, +1,\ 0$

NCO^-; $-2,\ 0, +1$

CNO^-; $-3, +1, +1$

(III) $-X\equiv\equiv\equiv Y-Z---$

N_3^-; $0, +1, -2$

NCO^-; $0,\ 0, -1$

CNO^-; $-1, +1, -1$

(IV) $\overset{x}{\underset{x}{X}}\ X\overset{0}{\underset{0}{=}}\ Y\overset{x}{\underset{x}{=}} Z\overset{0}{\underset{0}{=}}$

N_3^-; $-1, +1, -1$

NCO^-; $-1,\ 0,\ 0$

CNO^-; $-2, +1,\ 0$

(V) $\overset{x}{=} X\overset{0}{=}\ Y\overset{x}{=} Z\overset{0}{=}$

N_3^-; $-1\frac{1}{2}, +1, -\frac{1}{2}$

NCO^-; $-1\frac{1}{2},\ 0, +\frac{1}{2}$

CNO^-; $-2\frac{1}{2}, +1, -1\frac{1}{2}$

(VI) $\overset{x}{=} X \overset{0}{=} Y \overset{x}{=} Z\overset{0}{=}$

N_3^-; $-\frac{1}{2},\ +1, -1\frac{1}{2}$

NCO^-; $-\frac{1}{2},\ 0, -1\frac{1}{2}$

CNO^-; $-1\frac{1}{2}, +1, -\frac{1}{2}$

The formal charges on each atom can be easily computed and these are shown below each structure for N_3^-, NCO^- and CNO^-. Ideal and hence stable structures are those in which the charges are symmetrically disposed and in which the charges on N are from $-\frac{1}{2}$ to $+1$, on O from $-\frac{1}{2}$ to $+\frac{1}{2}$ and on C around $-\frac{1}{2}$. It can be seen that although N_3^- has two structures with symmetrical charges, the negative charges on the end N atoms are excessive, while NCO^- has one structure with both a symmetrical distribution and ideal negative charges on the end N and O atoms. The structures for CNO^- have the most asymmetric charge distributions. There, however, is one structure for CNO^- which is symmetric with respect to charge but the end atoms have excessive charge. These considerations clearly show that NCO^- would be the most stable and CNO^- probably the most unstable.

N_3^- and CNO^- can exist in a metastable state in ionic lattices. Among the azides, the anion is essentially unperturbed in the alkali metal salts but in the more complex heavy metal salts increasing perturbation of the anion occurs which is reflected in the asymmetric intraionic distances of the divalent salts in particular. This may be one of the reasons why the heavy metal salts are unstable with respect to the alkali metal azides. It is therefore pertinent to note that among the divalent azides the thermal sensitivity increases with the increasing asymmetry of the azide ions which increases in the order $BaN_6 < PbN_6 < CuN_6$ (cf. Table 2). Electron microscopic observations on thallous azide crystals have shown that the cubic form of the salt is relatively stable compared with the low temperature orthorhombic form (85). This is probably associated with the existence of asymmetric azide ions in the latter polymorph (cf.) Section II D).

All the metal fulminates are expected to be unstable because of the extreme charge asymmetry of the fulminate ion. The fulminates decompose with extensive polymerization (86) which is understandable in terms of the large dipole moment expected for the anion (*cf.* Tables 12 and 13). Differences in thermal sensitivity also exist among polymorphs of the fulminate salts. *Britton* and *Dunitz* (*31*) point out, for example, that the rhombohedral form of silver fulminate which has bent fulminate ions is probably unstable with respect to the orthorhombic form which has linear anions.

Pseudohalide crystals like the fulminates with asymmetric ions can be piezoelectric, while crystals which are sensitive to electric fields probably have ferroelectric phases. Electric fields generated on the application of thermal shock and pressure on piezoelectric crystals can cause electrical breakdown which readily leads to the formation of localized "hot spots" and the initiation of an explosive process. These ideas would need detailed study and can provide insights into the behaviour of energy rich crystals.

Acknowledgement. I would like to thank Drs. *H. D. Fair*, Jr., *H. J. Prask*, *J. Sharma* and *D. A. Wiegand* (Explosives Laboratory, FRL), and Dr. *S. K. Deb* (Cyanamid, Stamford) for critical reading and suggestions regarding the manuscript. I would also like to acknowledge many discussions on the lattice dynamics of the metal pseudohalides with Dr. *S. Trevino* (Explosives Laboratory, FRL) and Prof. *S. S. Mitra* (University of Rhode Island) and their permission to quote and depict (Fig. 6) unpublished results.

References

1. *Birkenbach, L., Kellermann, K.:* Chem. Ber. *58*, 876 (1925).
2. *Yoffe, A. D.:* Developments in Inorganic Nitrogen Chemistry (Ed. *Colburn*), Chap 2, Elservier (1966).
3. *Winnewisser, M., Cook, R. L.:* J. Chem. Phys. *41*, 855 (1964).
4. *Herzberg, G., Reid, C.:* Disc. Farad. Soc. *9*, 92 (1950).
5. *Reid, C.:* J. Chem. Phys. *18*, 1512 (1950).
6. *Winnewisser, M., Bodenseh, H. K.:* Z. Naturforsch. *22a*, 1724 (1967).
7. *Von Dohlen, W. C., Carpenter, G. B.:* Acta Cryst. *8*, 646 (1955).
8. *Kemmey, P. J., Mattern, P. L., Bartram, R. H.,* to be published.
9. *Pringle, G. E., Noakes, D. E.:* Acta Cryst. *B 24*, 262 (1968).
10. *Akers, D., Peterson, S. W., Willett, R. D.:* Acta Cryst. *B 24*, 1125 (1968).
11. *Wagner, E. L.:* J. Chem. Phys. *43*, 2728 (1965).
12. *Iqbal, Z., Yoffe, A. D.:* Proc. Roy. Soc. *A 302*, 35 (1967).
13. *Waddington, T. C.:* J. Chem. Soc., 2499 (1959).
14. *Pistorius, C. W. F. T., Boeyens, J. C. A.:* J. Chem. Phys. *48*, 1018 (1968).
15. *Pauling, L., Hendricks, S. B.:* J. Am. Chem. Soc. *47*, 2904 (1925).
16. *Gray, P.:* Quart. Rev. *27*, 441 (1963).
17. *Mueller, H. J., Joebstl, J. A.:* Zeit. für Krist. *121*, 5 (1965).

18. *Waddington, T. C.:* Ph. D. dissertation, Cambridge (1955).
19. *Iqbal, Z.:* Optics Communications, **2**, 33 (1970).
20. *Klug, H. P.:* Z. Krist. **85**, 214 (1933).
21. *Yamada, Y., Watanabe, T.:* Bull. Chem. Soc. Japan **36**, 1032 (1963).
22. *Frevel, L. K.:* Z. Krist. *A* **94**, 197 (1936).
23. *Zvonkova, Z. V., Zhadanov, G. S.:* Zh. Fiz. Khim. **23**, 1495 (1949).
24. *Llewellyn, F. J., Whitmore, F. E.:* J. Chem. Soc. London 881 (1947).
25. *Choi, C. S.:* Acta Cryst. *B* **25**, 2638 (1969).
26. *Iqbal, Z., Brown, C. W., Mitra, S. S.:* J. Chem. Phys. **52**, 4867 (1970).
27. *Handbook of Chem. & Phys.:* 43rd Ed. Chem. Rubber Publishing Co., Cleveland Ohio.
28. *Wilsdorf, H.:* Acta Cryst. **1**, 115 (1948).
29. *Pfeiffer, H. G.:* Ph. D. dissertation, California Inst. of Tech. (1948).
30. *Britton, D., Dunitz, J. D.:* Acta Cryst. **18**, 424 (1965).
    — Acta Cryst. **19**, 662 (1965).
32. *Lindqvist, I.:* Acta Cryst. **10**, 29 (1957).
33. *Choi, C. S., Boutin, H.:* Acta Cryst. *B* **25**, 982 (1969).
34. *Azaroff, L. V.:* Krist. **107**, 362 (1956).
35. *Söderquist, R.:* Acta Cryst. *B* **24**, 450 (1968).
36. *Suzuki, A.:* J. Industr. Explosives Soc. Japan, **14**, 142 (1953).
37. a) *Bradley, R. S., Grace, J. D., Munro, D. C.:* Z. Krist. **120**, 349 (1964).
    b) *Mitra, S. S.:* Solid State Phys. **13**, 1 (1962).
38. *Haase, O.:* private communication (1970).
39. *Pistorius, C. W. F. T.:* J. Chem. Phys. **51**, 2604 (1969).
40. *Iqbal, Z., Malhotra, M. L.:* to be published.
41. — J. Mol. Structure, **7**, 137 (1971).
42. *Gray, P., Waddington, T. C.:* Trans. Farad. Soc. **53**, 901 (1957).
43. *Maki, A., Decius, J. C.:* J. Chem. Phys. **31**, 772 (1959).
44. *Jones, L. H.:* J. Chem. Phys. **25**, 1069 (1956).
45. *Iqbal, Z.:* Trans. Farad. Soc., to be published.
46. *Bryant, J. I., Turrell, G. C.:* J. Chem. Phys. **37**, 1069 (1962).
47. *Dows, D. A., Pimentel, G. C.:* J. Chem. Phys. **23**, 1258 (1955).
48. *Reid, C.:* J. Chem. Phys. **18**, 1512 (1950).
49. *Beck, W., Fedl, K.:* Angew. Chem. (Intl. Ed.) **5**, 722 (1966).
50. *Orville-Thomas, W. J.:* Trans. Farad. Soc. **49**, 855 (1953).
51. More recent results on HN_3 by *Thompson, W. T., Fletcher, W. H.,* Spectrochim. Acta **22**, 1907 (1966).
52. *Iqbal, Z.:* 24 Symposium on Mol. Struct. and Spectroscopy, T9 (1969).
53. *Trevino, S., Mical, R. D.:* unpublished (1970).
54. *Malhotra, M. L., Moller, K. D., Iqbal, Z.:* Phys. Letters **31** *A* 73 (1970).
55. *Iqbal, Z.:* J. Chem. Phys., **53**, 3763 (1970).
56. —, *Malhotra, M. L.:* Spectrochim. Acta, **27** *A*, 441 (1971).
57. —, *Garrett, W., Brown, C. W., Mitra, S. S.:* J. Chem. Phys., to be published.
58. *Boutin, H., Trevino, S., Prask, H. J.:* Chem. Phys. **45**, 401 (1966) and unpublished results.
59. a) *Iqbal, Z.:* Second International Conference on Raman Spectroscopy, Oxford (1970)
    b) *Weir, C. E., Block, S., Piermarini, G. J.:* J. Chem. Phys. **53**, 4265 (1970).
60. *Bryant, J. I.:* J. Chem. Phys. **45**, 689 (1966).
61. *Savoie, R., Pezolet, M.:* Canad. J. Chem. **45**, 1677 (1967).
62. *Bowden, F. P., Yoffe, A. D.:* "Fast Reactions in Solids" Academic Press (1958).
63. *Ziman, J.:* "Phonons and Electrons", Oxford Univ. Press (1960).

64. *Gray, P., Waddington, T. C.:* Proc. Roy. Soc. *A 235,* 481 (1956).
65. *Gora, T.:* J. Phys. Chem. Solids, *32,* 529 (1971).
66. *Mitra, S. S.:* private communication (1970).
67. *Waddington, T. C.:* "Advances in Inorganic and Radiochemistry", Emeleus and Sharpe (Ed.), Acad. Press, Vol 1, p 157 (1959).
68. *Walsh, H. D.:* J. Chem. Soc. 2266 (1953).
69. *Scherr, V., McDonald, J. R.,* and *McGlynn, S. P.* (unpublished), Quantum Chemistry Program Exchange, Indiana State Univ., Bloomington, Indiana (QCPE No. 88).
70. *McDonald, J. R., Scherr, V. M., McGlynn, S. P.:* J. Chem. Phys. *51,* 1723 (1969).
71. *Rabalais, J. W., McDonald, J. R., McGlynn, S. P.:* J. Chem. Phys. *51,* 5103 (1969).
72. *McDonald, J. R., Rabalais, J. W., McGlynn, S. P.:* J. Chem. Phys. *52,* 1332 (1970).
73. *Clementi, E., McGlean, A. D.:* J. Chem. Phys. *39,* 323 (1963).
74. *Pererimhoff, S. D., Buenker, R. J.:* J. Chem. Phys. *47,* 1953 (1967).
75. *Siegbahn, K. et al:* Nova Acta R. Soc. Scient. Upsat. Ser IV, *20,* p. 111 (1967).
76. *Deb, S. K.:* J. Chem. Phys. *35,* 2122 (1961).
77. *Iqbal, Z.:* unpublished (1970).
78. *Deb, S. K.:* unpublished work, National Res. Council, Canada Private communication from *Yoffe, A. D.*
79. *Sharma, J., Fair, H. D.:* Bull. Am. Phys. Soc. *15,* 387 (1970).
80. *Fair, H. D., Forsyth, A. C.:* J. Phys. Chem. Solids, *30,* 2559 (1969)
81. *Deb, S. K. Yoffe, A. D.:* Proc. Roy. Soc. *A 235,* 106 (1960).
82. *McClaren, A. C., Rogers, G. T.:* Proc. Soc., *A 240,* 484 (1957).
83. *Bowden, F. P., Williams, H. T.:* Proc. Roy. Soc. *A 208,* 176 (1951).
84. *Linnett, J. W.:* "The Electronic Structure of Molecules — A New Approach", Metheuen p. 65 (1964).
85. *Iqbal, Z.:* Ph. D. dissertation, Cambridge, appendix 2 (1966).
86. *Boddington, T., Iqbal, Z.:* Trans. Farad. Soc. *65,* 509 (1969).

Received April 5, 1971

Spectra and Bonding in Metal Carbonyls, Part A: Bonding

Dr. P. S. Braterman

Department of Chemistry, University of Glasgow, Glasgow, W. 2., Great Britain

Table of Contents

I. General

Properties of compounds containing a CO group directly bound to a transition metal have been extensively reviewed in recent years (1). This work is therefore confined in Part A to a discussion of the nature of the bonds in metal carbonyls, especially metal-carbon and metal-metal bonds, and in Part B to the influence of these bonds on physical and spectroscopic properties. Other bonds will be discussed incidentally, because of their influence on coordinated CO and the correlative possibility of using coordinated CO as a probe into the bonding of the molecule as a whole.

The literature coverage is illustrative rather than exhaustive. Increasing knowledge about metal carbonyls has not led to increasing agreement, and the general positions are those of the author at the time of writing.

II. Qualitative Theoretical Treatment

II.1. Introduction

Since there is no simple relation between bonds and eigen-functions in a many-atom molecule, the discussion will generally use a molecular orbital rather than a valence bond approach. The molecular orbitals used to rationalise the bonding are linear combinations of atomic orbitals, this arbitrarily limited basis set being easy to visualise. We shall occasionally speak of altering the size and shape of these orbitals; this is equivalent to using an expanded basis set.

II.2. The Conventional Picture

It has been usual for many years (2) to regard the metal-carbon bond in carbonyls as possessing partial double-bond character. This can be represented in valence bond language as

$$\{M \leftarrow : C \equiv O \longleftrightarrow M = C = O\}$$

The simplest molecular orbital approach elaborates on this ideal, analysing the bond into two parts:

a) a 'forward' carbon-metal σ-bond

b) 'backward' metal-carbonyl π-bonds. There are two of these per CO, in two (generally arbitrary) mutually perpendicular planes. It is in a sense more correct to talk of a partial triple bond, rather than a partial double bond. These contributions to the bonding are shown in Fig. 1.

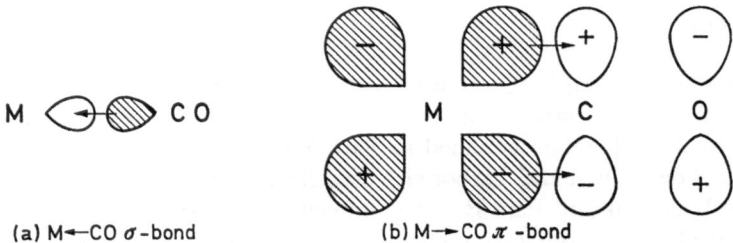

(a) M←CO σ-bond   (b) M→CO π -bond

Fig. 1. (a) M← CO σ-bond (b) M→CO π-bond

The forward bond uses the carbon lone pair, and a vacant metal orbital. It is characteristic of ligands in general. The ligand is behaving as a nucleophile, or Lewis base. This alone could not explain the formation of metal carbonyls. CO itself is a very poor base, coordinating (apart from transition metals) only to the BH_3 grouping[1]).

The 'back-bonding' uses metal orbitals conventionally regarded as metal nd-orbitals, though in environments of low symmetry they may acquire considerable $(n + 1)$ s, p character, and the vacant CO π^* orbitals. These must be concentrated mainly on the carbon atom (since the π-orbital is concentrated (4) on the oxygen atom), and overlap appreciably (5) with the metal d-orbitals. Such back-bonds are characteristic of unsaturated ligands, and of others which have low-lying empty orbitals of correct symmetry (such as phosphines, which have empty 3 d orbitals on the P atom). This picture explains the occurrence of the metal carbonyls; the fact that they occur mainly with metals in low oxidation states, and with several full d-orbitals; and (as discussed in Part B of this review) the lowering of C—O stretching frequency on coordination and the occurrence of CO (and other π-accepting ligands) high in the spectrochemical series (6).

The role of the full CO π-orbitals is neglected in this picture. This may be justified (except possibly for carbonyl-containing cations) on energetic grounds, Moreover, the π-orbital is localised mainly on the oxygen atom, distant from the metal, while the π^* is localised mainly on the carbon. The electrostatic field of the metal can in principle cause mixing of π (CO) with π^*(CO), but there is (III(2) below) reason to believe that this is unimportant.

II.3. Molecular Orbital Theory: Octahedral Case

The simple picture of the previous section is obviously incomplete. It discusses each CO group in isolation as if it had a monopoly of the metal. It is more usual to find several CO groups attached to the metal, and even where this is not so, other ligands are present.

The molecular orbital theory of complexes (6), developed out of crystal field theory (7) by the inclusion of both σ- and π-covalent interactions, has been successfully applied (8—13) to the octahedral (14, 15) carbonyls of the d^6 metals (V^{-1}, Cr^0, Mn^+ etc.). The qualitative discussion of this octahedral case is particularly simple because of a clean-cut separation (within the single-configuration approximation) of σ- and π-systems.

The σ-bonding scheme gives the familiar splitting of the metal nd-orbitals into t_{2g} (non-bonding) and e_g (σ-antibonding) sets, while the ligand lone pairs occupy bonding orbitals and the $(n + 1)s,p$ become σ-antibonding.

---

[1]) BH_3 is not only an unusually good σ-acceptor, but also, in this case, a hyperconjugated π-donor (3).

To discuss the π-bonding, the CO π^* orbitals may be combined to span the four triply degenerate representations t_{1g}, t_{2g}, t_{1u}, t_{2u} of O_h. Since orbitals must belong to the same symmetry class in order to mix, the t_{1g}, t_{2u} are non-bonding as regards the metal.

The t_{2g} interacts with the metal nd electrons, while the t_{1u} interacts with the vacant $(n+1)p$ σ-antibonding set.

The result is shown in Fig. 2. The π^* set are split into three main energetic groupings, the ligand field splitting, Δ, is increased, the complex is stabilised as the filled t_{2g} levels are lowered in energy, and the CO bond-order is decreased by an amount dependent on the degree of mixing of t_{2g} (nd) and t_{2g} (π^*). The CO 'valence bond order' will fall from the value of 3 assigned to free CO, to 2.75 if there is equal mixing of nd and π^* in the t_{2g} (2) level of Fig. 2. Energy considerations (8) suggest that the d-orbitals of a free metal atom in a (d^n) valence configuration are lower-lying than the CO π^* orbitals. The degree of back-bonding will then be less than this, and thus 2.75 may be taken as a plausible lower limit. The extreme case of complete back-bonding leads to an absolute lower limit of 2.5. The mixing of CO (π) with nd will also give t_{2g} (2) some CO (π) character. This will not reduce the CO bond order, since t_{2g} (2) is full. Mixing of CO (π) into the vacant t_{2g} (3) will reduce it, but this effect is second-order and presumably slight.

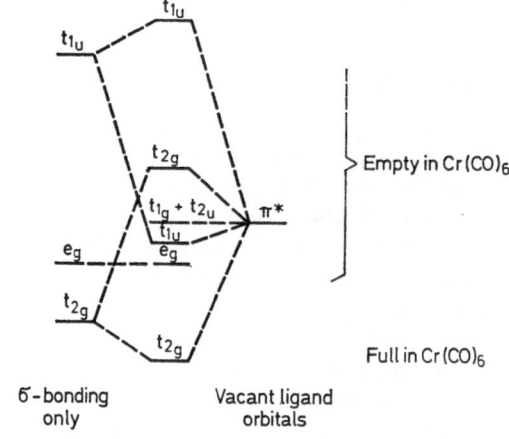

Fig. 2. Effects of π-bonding in an octahedral complex

II.4. Book-Keeping: Use of Orbitals

It is often instructive to count the number of valence-electrons associated with a metal ion. This is done (16, 17) by adding on to the number of outer electrons in the free ion the number of ligand electrons used in metal-ligand bonding. Thus each covalent bond (including metal-metal bonds) contributes one electron, each coordinated ligand (including coordinated double bonds) two, π-allyl three, dienes four, etc. In most known CO-containing complexes as in many others this number is 18 (*18*), and such complexes are said to obey the 18-electron rule.

This rule is, from a present-day standpoint, a rule of orbital use. Corresponding to any of the nd $(n + 1)$ s, p there will exist in the complex *either* a non-σ-bonding orbital *or* a σ-bonding-antibonding pair. (Weakly antibonding levels, such as d (t_2) in Ni $(CO)_4$, are treated as non-bonding[2]) The rule will then be obeyed whenever the σ-bonding and σ-non-bonding orbitals are all full and the σ-antibonding orbitals are all empty.

The use of this rule is restricted to strong-field ligands. The lowest σ^* orbital must be so high in energy that were it to be populated reaction would ensue. The σ-non-bonding orbitals must be stable enough to be filled, without these highest-energy electrons being transferred to the environment. For this to happen, the σ-non-bonding orbitals must be stabilised either by positive charge (as in Co^{3+} complexes) or by π-bonding (as in $Cr(CO)_6$).

Thus the rule is irrelevant (even when, as in $Fe(H_2O)_6^{++}$, accidentally obeyed) for all high-spin complexes, but is useful for carbonyls and other complexes in which back-bonding is important. It shows that a certain number of metal orbitals are used, but does not show whether the resultant molecular orbital is concentrated mainly on the ligands or on the metal. For example, co-ordinated hydrogen increases the electron count by one, whether described as $M^- \rightarrow H^+$ (one for charge, none for bond), $M-H$ (one for bond), or $M^+ \leftarrow H^-$ (-1 for charge, 2 for bond). The usefulness of the rule is thus, fortunately, independent of detailed assumptions concerning the actual electron distribution in the complex.

The rule has three main uses.

i) It draws attention to possible analogies between superficially dissimilar species (e.g. $C_5H_5Cr(CO)_3^-$ and $Co(CO)_4^-$).

ii) It rationalises the structures of a very large number of highly complicated substances, and can to some extent be used in prediction.

---

[2]) This procedure may be rationalised by taking suitable hybrids of nd orbitals and $(n + 1)$ s, p orbitals *of the same symmetry class* (which is possible in D_{3h} and T_d, but not in O_h), so that these hybrids are σ-non-bonding, and placing the 'd-electrons' in these. This is physically most plausible when the metal carries little positive charge, and is not at the very end of the transition series, otherwise the process will require a high promotion energy. This could be why Fe^0 generally obeys the rule, while Ni^{++}, even with ligands such as CN^-, generally does not, and Cu(I), Hg(II) are commonly 2- or 3-co-ordinate even with π-accepting ligands.

iii) Deviations from the rule, within a class of compounds to which it is generally applicable, raise (but do not answer) interesting questions about the bonding and reactivity.

For example, $V(CO)_6$ has 17 outer electrons. This shows that some orbital (presumably one of $t_{2g}(2)$ of Fig. 2) is only half-full, and gives a rationalisation for the easy reduction to $V(CO)_6^-$. $Ni(CN)_4^{2-}$ is two electrons short because one orbital (the $4p_z$ in the usual m.o. treatment (19)) is not used. This indicates the possibility of further coordination (as in $Ni(CN)_5^{3-}$). Nickelocene $Ni(C_5H_5)_2$ has two electrons in excess of the favoured number; these are in $3d$ antibonding orbitals (20), and explain the high reactivity of this species (21), in contrast to ferrocene.

II.5. Book-Keeping: Use of Electrons

The 18-electron rule describes the use of orbitals. This suggests a possible complementary rule for the use of electrons, namely that a role in bonding should be found for as many of the outer electrons of the metal as possible (22).

Thus in $Cr(CO)_6$ the only such outer electrons are used in π-bonding. The same is true of the d-electrons in $Fe(CO)_5$ and $Ni(CO)_4$. In $C_6H_6Cr(CO)_3$, which may be treated as pseudo-octahedral, the $d(xy)$ and $d(x^2-y^2)$ electrons of the metal combine with the π^* orbitals of both CO and C_6H_6, while $d(z^2)$ (which points along the 3-fold axis) donates exclusively to the former.

The rule fails trivially for all complexes of saturated ligands (PH_3 may be considered as unsaturated because of its empty $3d$ orbitals), even where the 18-electron rule is obeyed, since there is no possibility of back-bonding in these cases. It also fails for $(C_6H_6)_2Cr$, where the $d(xy, x^2-y^2)$ are used in back-bonding, but the $d(z^2)$ (or $d(z^2)-s$ hybrid) is either totally non-bonding or else feebly antibonding to the lowest aromatic molecular orbital of the rings.

The failure is even more extreme with $Ni(\pi\text{-allyl})_2$. Here each ligand provides three orbitals outside the carbon-carbon σ-skeleton. So (unless we are to say that carbon-carbon σ^* orbitals are populated) at most 6 m.o.'s, or twelve electrons, can be localised on the ligands or used in bonding to them. But there are 16 "outer" electrons. At least four of these must lack a bonding role[3].

The rule of electron use is evidently of much narrower application than the rule of orbital use (18-electron rule). It seems to hold only in the

---

[3] In $Ni(CN)_4^{2-}$, on the other hand, all but two "metal" electrons ($d(z^2)$) are involved in back-bonding. Here electron counting suggests an analogy between two compounds, but the criterion of electron use shows the limitations of this analogy.

presence of several π-accepting groups, when it is trivial. This asymmetry between electron use and orbital use means that transition metals, even in their lowest oxidation states, are better electrophiles than nucleophiles. Nonetheless, it may be useful to ask how many outer electrons lack a bonding role, and whether this can be related to chemical reactivity.

II.6. Tetrahedral and Lower Symmetries

The naive application of the picture of Fig. 1 to $Ni(CO)_4$, for example, is totally misleading. It is assumed that the metal can present each CO with one empty pure σ-orbital and two full π-orbitals. Thus each CO in turn is singled out for special attention, and a description offered which cannot be applied at the same time to the other three. The conventional picture has, in fact, chosen an axis of quantisation along one of the metal-carbon bonds. This is legitimate for the octahedral case, but not for tetrahedra. The defect is readily removed by a qualitative molecular orbital approach, illustrated in Fig. 3.

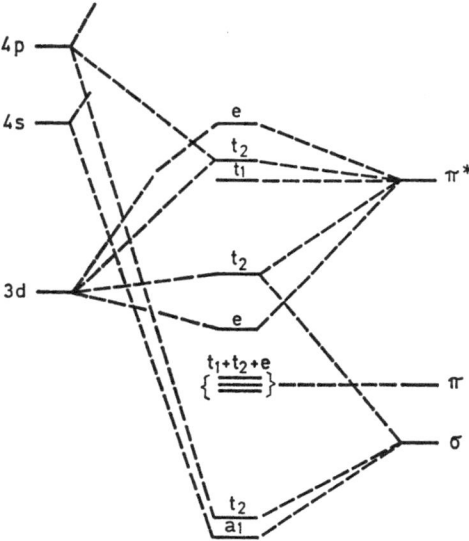

Fig. 3. Qualitative m. o. diagram for a tetrahedral metal carbonyl (ligand 4 σ and 5 σ not shown separately)

There is a further, more serious, complication. The CO σ-lone pairs, and the π and π^* orbitals, all span t_2. This means that there can be interaction between the σ-lone pair of each CO and the empty π^* orbitals of the others. This effect has been overdramatised by the suggestion (22) that where such σ-π^* interaction is possible the role of the metal is largely to "short-circuit" carbonyl orbitals from different CO groups into each other.

The simpler view can be defended on three counts. Firstly, if this "short-circuiting" were energetically favourable, one would expect CO groups to add to each other in the absence of a metal, and spontaneously to form clusters. Secondly, if the role of the metal is as suggested, positive charge on the metal should attract both of the sets of orbitals to be short-circuited towards a common centre, and enhance the effect. Then we would expect tetrahedral carbonyls of Cu^+ and Zn^{++}. Moreover, the higher the charge on the metal, the more short-circuiting could occur, the greater the reduction in CO bond order and the lower the CO stretching frequency.

Finally, the greatest loss of short-circuiting from a molecule such as $Ni(CO)_4$ should occur with the loss of the first CO group, which we would then expect to be the most strongly bound

$$(D[(OC)_3Ni{-}CO] > \bar{D}[Ni{-}(CO)_4]).$$

The exact opposite is found. Compounds of alleged formula $(CO)_n$ are stabilised by hydration to $[C(OH)_2]_n$ (23), and decompose on removal of the water (24). Cu^+ has a very limited capacity for coordinating CO (25), while Zn^{++} forms no carbonyl complexes, but $Co(CO_4)^-$ and $Fe(CO)_4^{2-}$ are familiar. The CO stretching frequency falls, and therefore the population of CO π^* presumably increases, along the series Ni^0, Co^{-1}, Fe^{-2} (26). The first CO group is much more readily lost from $Ni(CO)_4$ than average bond-strengths would suggest ($D = 13$ kcal, $\bar{D} = 45.7$ kcal/mole (27)) and this result may be rationalised by considering that $M \rightarrow L$ π-bonding stabilises the $Ni(CO)_3$ fragment (28). All these facts are in accordance with the view of Fig. 3, and directly contrary to the predictions of the "short circuiting" hypothesis. Moreover, were such "short circuiting" energetically significant one would expect symmetrical molecules, such as $Cr(CO)_6$, to show a spontaneous distortion so as to allow this short-circuiting to take place.

This last consideration applies forcibly to some recent discussions of the cis-$M(CO)_3$ grouping, as found in $C_6H_6Cr(CO)_3$. Here we can distinguish between 'axial' and 'tangential' π^*-orbitals, the former lying in the C_{3v} reflection planes, the latter at right angles to them. π^*-axial transforms in the same way as σ, and short-circuiting is allowed not only between σ of one group and π^* of another, but between σ and π^* of the same CO. A thoroughgoing short-circuit theory combines the $\sigma(CO)$ and π^* (axial) set, so that the only metal$-$CO bonding is the back-bonding between the d-orbitals and the π^*-tangential set. We can then expect the two different sorts of π^* to be filled to different extents, and predict appreciable non-linearity in the M—C—O groups. However, the observed distortion hardly ever exceeds 5^0 (29, 30)[4].

One case of lowered symmetry of particular interest is that which arises when one CO group in an octahedron is replaced to give a derivative $M(CO)_5X$. If the group X is less π-accepting than CO (this is probably true of all common groups except, perhaps, PF_3 (32)) then the CO group trans to the group X will be uniquely bonded. This CO group uses the $d(xz, yz)$ orbitals (taking X to lie on the z-axis) each of which is shared between three carbonyls. Each of the radial carbonyls uses only one of these, while all four of them compete for the remaining, $d(xy)$, orbital. Moreover, if the group X is a π-donor (e.g. Cl), or a π-non-bonding group capable of repelling electrons (e.g. CH_3), we expect the $d(xz, yz)$ orbitals to be distorted away from the group X, by admixture of a small amount of $(n+1)p(x, y)$. This distortion does not affect the bonding between the modified d-orbitals and the four radial carbonyls, but increases that between these orbitals and π^*(axial). For both these reasons the axial carbonyl will be uniquely strongly bonded. How strongly, and how much this will upset the simple molecular orbital scheme of Fig. 2, is one of the main topics of part B.

II.7. Metal-Metal Bonds, and Bridging Groups

$Mn_2(CO)_{10}$ provides an example of the simplest sort of metal-metal bond. The 18-electron rule is obeyed; the electron use rule is satisfied (by using one electron from each metal atom in σ-bond formation, and allowing the others, as in systems $M(CO)_5X$, to π-bond to the carbonyls), and a molecular orbital approach would simply regard the σ-bond as, in effect, a ligand and treat the metal as distorted octahedral diamagnetic Mn^+ (t_{2g}^6).

---

[4] *Kettle* (30) gives a further, more subtle, argument in favour of non-linearity. He shows that the sum of exchange integrals between π^* (axial) and the d-orbitals is different from the corresponding sum for π^* (tangential), although the sum of squares of exchange integrals is the same. Thus the energy of the two sets should be the same, but the amount of mixing of wave-functions should be different.

However, the chosen z-axis is C_3, not C_∞; xz, yz ($m_\ell = \pm 1$) and xy, x^2-y^2 ($m_\ell = \pm 2$) both span the representation E, and mix to an unknown extent. If forward-bonding is very much more important than back-bonding, as on the conventional approach, then $C_6H_6Cr(CO)_3$ is a pseudo-octahedron, the $Cr(CO)_3$ fragment is analogous to that in $Cr(CO)_6$, and no distortion is expected. This result is surprisingly close to what is found.

The root of the difficulty is (31) that all presuppositions about the basis set other than those required by symmetry are to some extent subjective, and that the results of calculations may in unfavourable cases merely expose the unforseen consequences of innocent-seeming, but arbitrary, assumptions.

Fe$_2$(CO)$_9$ raises a different set of problems. The structure (*33*) is shown in Fig. 4 (a), and the conventional bonding scheme (2a) in Fig. 4 (b). A single metal-metal bond is conventionally drawn so as to satisfy the 18-electron rule and to account for the observed diamagnetism, the short iron-iron distance and the acute FeCFe angle.

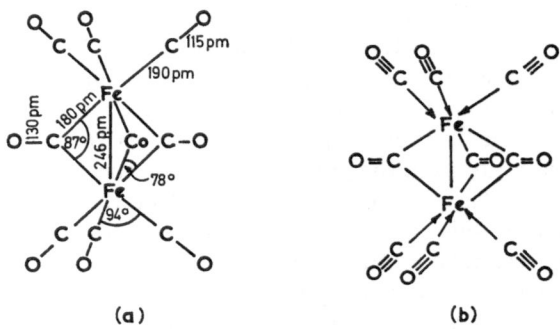

Fig. 4. (a) Structure of Fe$_2$(CO)$_9$ (after (*33*)); (b) Valence bonding scheme for Fe$_2$-(CO)$_9$

This approach has four main defects. Firstly, the iron atom is assigned an oxidation number of $+3$ ($+4$ if we include the metal-metal bond); this seems incompatible with the π-bonding requirements of the terminal CO's. Secondly, the CO stretching frequency in bridging carbonyls is around 1850 cm^{-1} (*34*), as against 1715 cm^{-1} for aldehydes and ketones (*35*), and 1630 for transition metal acyls (*36, 37*). Thirdly, no way of choosing the hybridisation at carbon can account for the fact that here, as in other bridged carbonyls, the MCM angle is in the range 80—85° (*38, 39*). Bent bonds or geometric strain seem unsatisfactory *ad hoc* explanations, in view of the widespread occurrence of these compounds. Finally, this scheme can only be extended to triply bridging CO (a feature common among polynuclear carbonyls (1*d*, 1*g*)) by a formulation such as

$$
\begin{array}{c}
\text{O}^- \\
| \\
\text{C} \\
M \diagup \,|\, \diagdown M \\
M
\end{array}
$$

which is inconsistent with the observed (ca. 1600 cm^{-1}) stretching frequencies.

A totally different description seems in order, using orbitals delocalised over a bridging carbon and the bridged metal atoms (*40, 1* (*g*)).

The bridging carbonyl is regarded as derived from free CO, and the metal is regarded as surrounded by six essentially d^2sp^3 hybrids pointing towards the vertices of a distorted octahedron[5]), the remaining d-orbitals being σ-non-bonding. This non-bonding set is analogous to t_{2g} of the true octahedral case. It may be analysed into totally symmetrical (a_1) and doubly degenerate (e) components in the C_{3v} symmetry around each metal atom.

The eight electrons of each iron atom can then be distributed as follows: 4 into the e (non-σ-bonding) set, one each into the three metal-bridging CO hybrid orbitals, and one into the a_1 (Fe—C σ-non-bonding) orbital. The two a_1 orbitals on the different iron atoms are then paired to give a σ-bond.

This bond is real, not formal. The a_1 orbital is $d(z^2)$ in C_{3v} and is therefore concentrated along the metal-metal axis. The metal-metal distance is short (246 pm)[6]).

The use of the remaining orbitals is shown in Fig. 5. We take the in-phase combination of the two iron σ-orbitals as empty (Fig. 5a) and let the resultant orbital, symmetrical (a') in the reflection place C_s, overlap the CO lone pair. The resulting three-centre bond is entirely analogous to the three-centre bonds in $Al_2(CH_3)_6$ or $[Be(CH_3)_2]_n$, in which the MCM angle is also anomalously small (*43, 44*).

One electron from each Fe is then placed in the out-of-phase combination (antisymmetric (a'') in C_s), which overlaps with the empty π^* orbital of the bridging CO group to give the three-centre π-bond of Fig. 5 (b).

This model differs from the conventional one in possessing an extra degree of freedom. The two conventional C—M bonds at each carbon can be taken in symmetric and antisymmetric combination to give the orbitals of Fig. 5 (a, b); but the conventional picture requires the weights of carbon and metal to be the same in the bonding as in the antibonding combination. This assumption is unnecessary and presumably false.

We may note that the rule of electron use is satisfied in this model by overlap of the e (σ-non-bonding) metal electrons with CO π^* orbitals. The 18-electron rule is satisfied if we maintain that each bridging CO donates two electrons to the Fe_2 unit and thus donates one to each Fe

---

[5]) The hybridisation is, of course, an artefact, and should strictly be replaced by an equivalent set (*41*) related to the C_{3v} symmetry. This is largely irrelevant to a discussion of ground state properties.

[6]) A more recent structure involving both bridged and unbridged Fe—Fe bonds is that of $Fe_3(CO)_{11}PPh_3$ (*42*). Here the bridged distance is 2.71 Å, while the distance between unbridged iron atoms in the same complex is 2.56 Å.

atom. The treatment of triply-bridging CO is similar (Figs. 5b, 5c); the a_1 combination of metal σ-orbitals overlaps the CO lone pair, while the e combinations are filled with back-donating 'metal' electrons.

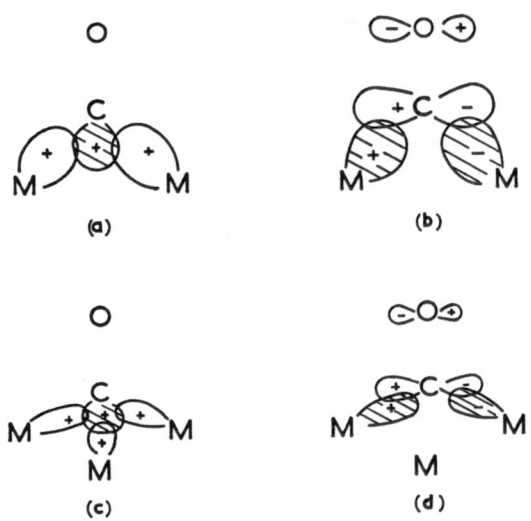

Fig. 5. Bonding of CO in bridging carbonyls: (a) forward donation to two metal atoms; (b) back-bonding from two metal atoms; (c) forward donation to three metal atoms; (d) back-donation from three metal atoms (one member of doubly degenerate set)

II.8. More Complicated Structures: The Question of "Bent Bonds"

The structure of $Co_2(CO)_8$ (Fig. 6(a)) in the solid is (39) surprisingly close to that of $Fe_2(CO)_9$. One bridging CO is missing in an otherwise similar structure. The 18-electron rule is satisfied once more by assuming a Co—Co bond. The nature of this bond is, however, a matter of current dispute. The question is of some importance since there exist a large number of compounds (e.g. $(OC)_3Fe[SMe]_2Fe(CO)_3$, (45), $(OC)_3Fe(S–S)$-$Fe(CO)_3$ (46), $(OC)_3Fe[PR_2]_2Fe(CO)_3$ (47) which are known (45, 46) or presumed (47) to have similar geometry to $Co_2(CO)_8$, and for which the electronic book-keeping and, presumably, (48), the nature of the bonding are closely related. [A bridging SMe group, for example, can be formally represented as — S(Me)+ —, and thus acts as a three-electron donor to the M_2 unit.]

There are two extreme positions which I shall call the 'straight bond' and 'bent bond' pictures. On the 'bent bond' theory (49), (Fig. 6(b)), six of the nine electrons of the Co atom fill the "t_{2g}" set of an idealised octahedron. Of the remainder, two are used in metal-bridging CO σ-bonding (this part of the model could be readily modified as suggested for $Fe_2(CO)_9$) and the remaining electron is directed towards that apex occupied in $Fe_2(CO)_9$ by a bridging CO, but vacant in this case. The in-phase combination of octahedral hybrids is taken, giving a bent σ-like bond.

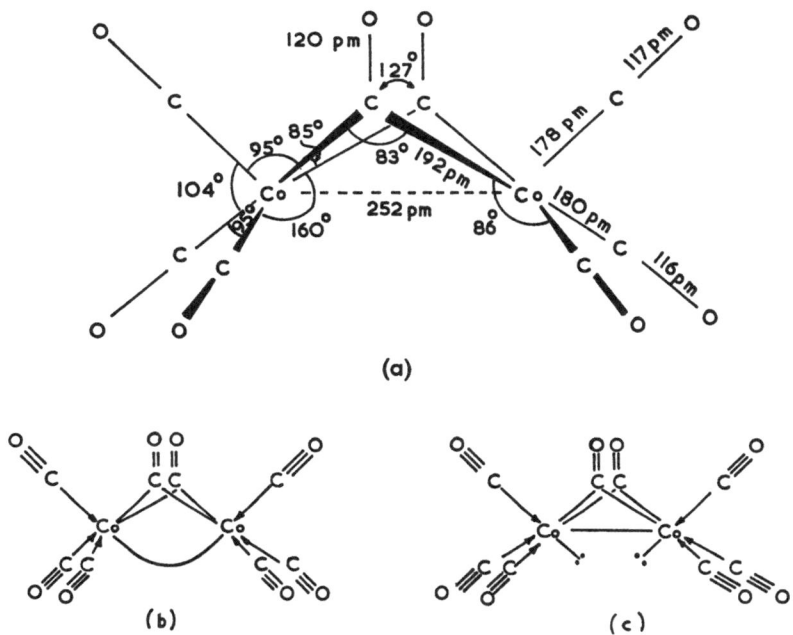

(a)

(b)　　　　　　　　　(c)

Fig. 6. Dicobalt octacarbonyl: (a) structure (after (39)); (b) bent bond description; (c) straight bond description

The 'straight bond' model (39) (Fig. 6(c)) treats $Co_2(CO)_8$ as more closely analogous to $Fe_2(CO)_9$. The 't_{2g}^5' configuration round each metal atom is retained, as is the straight metal-metal bond. The place of the missing CO group, with its lone pair and two metal electrons, is now occupied by two lone pairs, one on each metal atom. More precisely, both the in-phase and out-of-phase combinations involving this site are now full.

P. S. Braterman

In the notation of Fig. 7, the straight bond model requires an electronic configuration of type $(I)^2(II)^2(IV)^2$, while the bent bond model configuration is $(I)^2(II)^2(III)^2$. While there is no difference in symmetry type orbitals III and IV, the electron density distributions on these orbitals are quite different. It follows that the distinction between the two extremes is meaningful, although intermediate situations are possible in the low-symmetry cases under dispute.

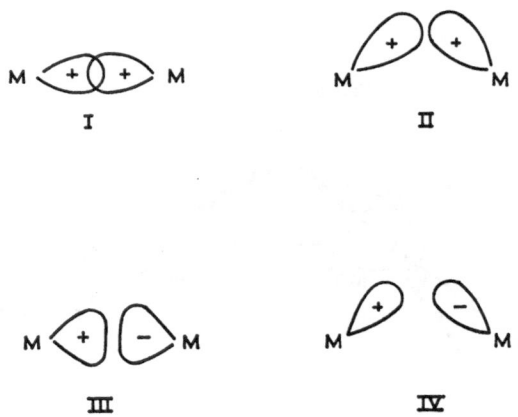

Fig. 7. Bent and straight bonds (bridging hydrogen may be present off the metal-metal axis)

A straight bond model is the only one applicable to such compounds as $(C_5H_5)_2Fe_2(CO)_4$ (38), $(OC)_4Cr[PMe_2]_2Cr(CO)_4$ and related compounds (50, 51), and $(ON)_2Fe(SEt)_2Fe(NO)_2$ (52). It follows that acute angles between metal-(bridging ligand) and metal-metal bonds are possible. On the other hand, a bent bond has been demonstrated in $[(Ph_3P)(NO)Ir]_2O$ (53)[7].

Bent bonds have been claimed in $(Rh(CO)_2Cl)_2$ (49) and in hydride bridged compounds (see e.g. (54)). Both these latter claims can be challenged. The dimeric units in $(Rh(CO)_2Cl]_2$ are puckered and stacked, in a way which no doubt allows some interaction between the filled $d(z^2)$ orbital of each Rh atom and the vacant $p(z)$ orbitals of its neighbours, but there is no reason to suppose the promotion of Rh to a $d(z^2)^1 p(z)^1$

[7]) A straight bond formalism for this compound would require the iridium atom to be related to a hypothetical T-shaped three-coordinate d^{10} system (just as, on a straight bond model, the cobalt atom in solid $Co_2(CO)_8$ is to a square pyramidal d^8 system). No such systems are known or expected.

configuration, as would be required for true covalent bonding. Bridging hydride no doubt forms a bent three-centre bond of type II (in the notation of Fig. 7), but this has nothing to do with the question of whether it is orbital (III) or (IV) that is occupied.

Despite the absence of any simple conclusive argument, the author feels drawn to the straight bond model of $Co_2(CO)_8$. In part, this is because of the analogy with those compounds where the presence of a straight bond has been established, and a dislike of assuming a non-bonded contact over a distance necessarily shorter than the total length of the bent bond[8]), but the most conclusive argument comes from the analogy (section II(9) below) between $Co_4(CO)_{12}$ and $Fe_4(CO)_{13}^{2-}$ on the one hand, and $Co_2(CO)_8$ on the other.

II.9. Polynuclear Compounds

The properties of polynuclear metal carbonyls have been reviewed (1 g).

Metal cluster compounds are expected to show a greater tendency than simple carbonyls to violate the 18-electron rule. A formal valence bond structure, involving localised metal-metal bonds, is likely to be a poor representation of the balance between several bonding and anti-bonding molecular orbitals (57). The coordination number will be large, as will the number of formally σ-antibonding orbitals. The breakdown of the rule in such circumstances (e.g. in $Rh_6(CO)_{16}$ (58) is not surprising. The problem is, rather, to explain why the rule is so well obeyed by large numbers of trinuclear and tetranuclear species, whether bridged (Fe_3-$(CO)_{12}$ (59), $[ReH(CO)_4]_3$ (60), $Ni_4(CO)_6[P(C_2H_4CN)_3]_4$ (61)) or unbridged ($Os_3(CO)_{12}$ (62), $Ir_4(CO)_{12}$ (63)).

Several compounds are known where one bridging group is shared by three metal atoms (e.g. $Cp_3Ni_3(CO)_2$, where each CO is linked to three nickels (64), and the tetrahedral species $[Mn(CO)_3SR]_4$ (65) and $[CpFe(CO)]_4$ (66)).

A detailed comparison of these last two species will bring out the differences between π-accepting and π-donating bridging groups. There is no metal-metal bonding in the manganese compound (67), and it follows that, if the 18-electron rule is to be obeyed, the RS^- ligand must be acting as a six-electron donor. Iron-iron bonding is present, however, and if we assume a normal σ-bond along each edge of the tetrahedron, the triply bridging CO group donates only two electrons, in accord with

---

[8]) Admittedly, different classes of orbital are involved, but it is relevant that the length of '$d-d$ bond' in $(OC)_4Mn[H][PMe_2]Mn(CO)_4$ (55), which can be considered strictly isoelectronic with $(OC)_4Cr[PMe_2]_2Cr(CO)_4$, is the same as that (56) of the 'hybrid-hybrid' bond in $Mn_2(CO)_{10}$.

theory of carbonyl bridges presented in II(8). (The reduction (68) of $Co_6(CO)_{15}^{2-}$ to $Co_6(CO)_{14}^{4-}$ confirms this two-electron character).

The 18-electron rule is obeyed by the vast majority of tetrahedral tetranuclear carbonyls of which the reviewer has knowledge. This could be because of the possibility of rearrangement[9] or because there is in this case (unlike the octahedral case discussed below) a fortuitous coincidence between the number of formal σ-bonds that a metal can make with its neighbours and the number of strongly overlapping orbitals, not pre-empted by other ligands, available for such bonding. (We have assumed π-overlap to have no structural consequences; this again seems fortuitously true of tetrahedra, though not of the triangular $Pt_3(COD)_3$ $(SnCl_3)_2$[10]) discussed below).

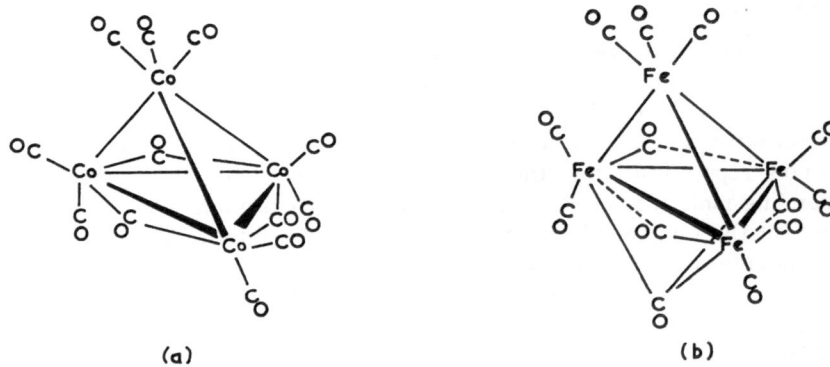

Fig. 8. (a) $Fe_4(CO)_{13}^{2-}$ (after (71)); (b) $Co_4(CO)_{12}$ (after (72))

It is instructive to compare $Fe_4(CO)_{13}^{2-}$ (71) and $Co_4(CO)_{12}$ (72) (Fig. 8). The apical iron atom in $Fe_4(CO)_{13}^{2-}$ may be assigned a formal negative charge, and is then isoelectronic with the apical cobalt atom in $Co_4(CO)_{12}$, or with any of the iridium atoms in $Ir_4(CO)_{12}$ (63). The basal iron atoms each have eight neighbours; the apical iron atom, two terminal CO groups, two edge-bridging CO groups, two basal iron atoms and the

---

[9] Thus $Re_4(CO)_{16}^{2-}$ (69), with a parallelogram of Re atoms bonded along the sides and one diagonal only, obeys the rule; hypothetical tetrahedral $[Re(CO)_4]_4^{2-}$ would have two excess electrons. Two fused M_3 triangles are also found in Pt_4-$(PPhMe_2)_4(CO)_5$ (70); in this latter case two of the Pt atoms have a 16-electron count, and a coordination number (excluding metal-metal bonds) of three (as in $Pt(PPh_3)_3$).

[10] COD = 1,5-cyclooctadiene.

face-bridging CO group. We can use octahedral 'hybrid orbitals' on the iron atoms for all their bonds except the bridged, basal-basal, iron-iron bonds (the edge-bridging CO's are in fact highly assymetrical, intermediate between bridging and terminal in character, but that does not affect the argument). The bridged metal-metal bonds are then built up from two 't_{2g}' orbitals and two electrons on each iron atom. The third 't_{2g}' orbital contains a non-σ-bonding pair. There are eighteen other bonds (six to each iron) and 36 remaining electrons (four from each iron, two from each of ten CO groups, three from basal-apical iron-iron bonds and one from the remaining negative charge). Thus the 18-electron rule is obeyed for each iron atom.

$Co_4(CO)_{12}$ may be described in exactly the same way, except that the site occupied in the iron compound by a face-bridging CO group, and thus by four 'metal' and two 'ligand' electrons, is now to be occupied by (symmetry-correct combinations of) three cobalt lone pairs. The basal cobalt atoms are five-coordinate (if as usual we discount bridged metal-metal bonds) with the sixth site of the coordination octahedron occupied by a lone pair and the 18-electron count achieved by straight, stereochemically inactive cobalt-cobalt bonds. This is *exactly* the situation required by the 'straight bond' description of $Co_2(CO)_8$ (II(8) above).

There are three situations in which the 18-electron rule is not followed by polynuclear metal carbonyls. The first, exemplified by $Pt_4(PPhMe_2)_4$ $(CO)_5$ (see above) is trivial; an atom which frequently fails to use all its orbitals in mononuclear complexes shows similar behaviour in a polynuclear species. More interesting are cases where the 18-electron rule describes an ideal bonding situation, from which some molecules deviate. Such molecules include $(C_5H_5)_3Ni_3(CO)_2$ (67) and $Co_3(CO)_9S$ (73). In each of these two cases, there is one more electron than the rule requires. The unusual electron count is not an inherent feature of the structure type, since the closely analogous species $(C_6H_6)_3Co_3(CO)_2^+$ (74) and $Co_3(CO)_9$-CMe (75) do fit the rule.

The truly antibonding nature of the highest orbitals involved has been demonstrated by a beautiful series of comparative studies. For example, six electrons in $(C_5H_5)_3Ni_3(CO)_2$ are σ-bonding and one is σ-antibonding. The Ni-Ni distance is 239 pm (64). In the closely related $(C_5H_5)_3Ni_3S_2$, there are six σ-bonding and five σ-antibonding electrons, and the Ni—Ni distance has increased to 280 pm (76). $Co_3(CO)_9S$, with one σ-antibonding electron, has an average Co—Co distance of 264 pm (72); in $Co_2Fe(CO)_9S$, which has one less electron but is otherwise similar, the average metal-metal distance is 255 pm (77).

There exists a third class of polynuclear compounds, including some polynuclear carbonyls, where the rule seems simply inapplicable. Either a σ-non-bonding molecular orbital is so high in energy as to be unpopulated, or a formally σ-antibonding orbital is so little destabilised, perhaps because of poor overlap, as to be virtually non-bonding, and thus full. The former situation obtains (79) in $Pt_3(COD)_3(SnCl_3)_2$ (78). Here the three platinums constitute a 44 electron system (3×10 (Pt) $+ 3 \times 4$ (COD) $+ 2 \times 1$ (SnCl$_3$)) where the rule requires 48[11]). The suggestion is that one doubly degenerate π-antibonding level, built out of $d(yz)$ orbitals (y is chosen perpendicular to the plane of the Pt$_3$ triangle, and z to point towards the triangle's centre) is empty. We then have some Pt—Pt π-bonding in addition to the σ bond which is all that the 18-electron rule requires; hence the Pt—Pt distance of 258 pm, as against an average of 276 pm for a normal σ-bond in $Pt_4(PPhMe_2)_4(CO)_5$.

The latter situation, in the author's opinion, obtains in the octahedral cluster compounds with total electron count 86: — $Rh_6(CO)_{16}$ (58), $Co_6(CO)_{16}$ (79), $Co_6(CO)_{15}^{2-}$ (90), $Co_6(CO)_{14}^{4-}$ (81), $Ru_6(CO)_{17}C$ (82) (treating the central carbon as a four-electron donor, as in $Fe_5(CO)_{15}C$ (83)), $Ru_{12}(CO)_{30}^{2-}$ (84) (counting the electron pair of the metal-metal bond joining the two Ru$_6$ octahedra as belonging to both) and $Ru_7(CO)_{16}^{3-}$ (85) (if the seventh, face-capping Ru atom makes normal σ-bonds to its three neighbouring Ru atoms, and has a count of 18). The expected count for an octahedral cluster is 84 ($n = 6$, $x = 12$) but no octahedral carbonyl clusters are known with this count ('$Ru_6(CO)_{18}$' having been re-characterised as $Ru_6(CO)_{17}C$ (85)). Evidently, these 86 electrons form some sort of closed shell.

A Hückel-type treatment of octahedral metal clusters was put forward some time ago (87) in order to account for the existence of the stable diamagnetic halide clusters $Mo_6Cl_8^{4+}$, $Ta_6Cl_{12}^{2+}$. This treatment regarded orbitals directly σ-bonding to halide ligands as energetically inaccessible (an assumption which could safely be carried over to the carbonyl clusters) but ignored all other ligand field effects, and assumed all available d-orbitals to have the same radial wave function. This treatment gives a wrong electron count for $Co_6(CO)_{14}^{4-}$, since it requires the last two electrons to be placed in a doubly degenerate orbital (in the solid salt $Co_6(CO)_{14}^{4-}$ is far from having perfect octahedral symmetry (81) but the actual S_6 symmetry does not affect the degeneracy in question)[12]). There has also been an ambitious attempt (88) to treat

---

[11] In general, for a system with n metal atoms and x metal-metal bonds, the rule requires a total of $18n - 2x$ electrons, since the two electrons of each metal-metal bond are counted twice.

[12] The present discussion assumes that the 86-electron clusters possess the same effective configuration, and that the diamagnetism experimentally demonstrated for some members of the class is common to all. This is admittedly going beyond the evidence, but the author cannot believe that the frequent recurrence of the 86-electron count is fortuitous.

$Rh_6(CO)_{16}$ by assuming that each symmetry combination of the empty (π^*) orbitals of the CO groups overlaps with one and only one full orbital in the molecule, which may or may not also be used in metal-metal or metal-ligand σ-bonding. The method leaves one π^* combination in $Co_6(CO)_{16}$ without a partner; in $Co_6(CO)_{14}^{4-}$, as it happens (*89*), it works perfectly. To this author, the treatment seems unsatisfactory; the rules require a short-circuiting mechanism (criticised in II(6) above); their precise application varies from structure to structure (so that the recurrence of the 86 count is fortuitous), and in any case the rules are not obeyed by simple carbonyls (e.g. $Cr(CO)_6$).

The simplest explanation of the 86-electron count is as follows (*1*(*g*), *80*, *40*): in $Co_6(CO)_{16}$, for example, the metal atoms are all four-coordinate in CO. This accounts for their 4 s,p orbitals, which go to make 24 full metal-carbon σ-bonds (including the 'e' face orbitals[13]) that overlap triply bridging CO). The $d(xy, x^2-y^2)$ orbitals (in the notation of Fig. 9) are not properly oriented for good metal-metal σ-overlap, and

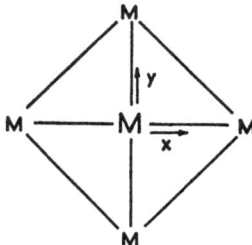

Fig. 9. Axis system for metal at apex of octahedral cluster; z-axis normal to plane of paper

are therefore treated as non-bonding, and all eight such orbitals are full. There remain the $d(z^2)$ and $d(xz, yz)$ orbitals, and 14 electrons. Taking combinations of these in O_h symmetry, we have

$$\Sigma d(z^2) \quad = A_{1g}(\text{bonding}) + E_g(\text{antibonding}) + T_{1u}(\text{antibonding})$$

$$\Sigma d(xz, yz) = T_{1u}(\text{bonding}) + T_{2g}(\text{bonding})$$

$$+ T_{2u}(\text{antibonding}) + T_{1g}(\text{antibonding}).$$

Thus we have constructed seven bonding electrons to accomodate the 14 electrons, and all bonding and non-bonding orbitals are full.

---

[13] In the C_{3v} symmetry of the M_3CO fragment

A similar argument can be constructed for all the octahedra (assuming only that the central carbon in $Ru_6(CO)_{17}C$ is satisfied by overlap with the $A_{1g}(d(z^2)$, and $T_{1u}(d(xz,yz))$ orbitals of the Ru_6 cluster). The orbitals used in constructing localised σ-bonding orbitals along the edges of the octahedron would be $d(z^2)$, $d(xy,yz)$ and $d(x^2-y^2)$; the twelve bonds then correlate with the bonding and non-bonding m.o's derived from these a.o's, and the 'extra' electron pair may be assigned to the $A_{2g}(x^2-y^2)$ orbitals, which is implicitly assumed empty by the valence bond treatment because of its *largely formal* anti-bonding character.

III. Quantitative Calculations

III.1. General Considerations

Both Hartree-Fock and Wolfsberg-Helmholz and related methods have been applied to metal carbonyl compounds. The Hartree-Fock method (90) treats each electron as moving in a potential field due to the nuclei and to the charge-density of all the other electrons. The interelectronic repulsion potential takes account of the tendency of electrons of parallel spin to avoid each other (exclusion), but not of correlation due to electron-electron repulsion independent of spin[14]). The field due to some assumed electron distribution is evaluated, and wave-functions calculated for motion in that field. When sufficiently close agreement between assumed and calculated wave-functions is finally obtained, the solution is said to be 'self-consistent'.

A completely *a priori* calculation of this type is impracticable for large molecules, and the result would in any case have to be transformed to some familiar basis before it generated any chemical insight. It is therefore customary (92) to restrict oneself to a basis set of, for example, atomic orbitals, so that each molecular wave-function is a 'linear combination of atomic orbitals.' The coefficients in the 'LCAO' expansion are then the variables, and these are adjusted at each stage to minimise the total energy. For large molecules, the problem is further simplified by designating certain inner electrons 'core electrons', and fixing their distribution to that in the free atom, or some other simple model system. The basis set are then made orthogonal to the core. Where there is negligible overlap between two parts of the molecule, their interaction is often ignored.

---

[14]) The chemically significant quantity that is being neglected is the 'extra correlation' present in the molecule in addition to that present in the separated components. This 'extra correlation' has been estimated for ethylene (91) as 7.7 e.v., which is about as much as the entire metal-carbon bonding energy in nickel carbonyl.

The meaning of the quantities evaluated in this way is quite clear. With each occupied one-electron wave-function is associated an energy ε which for closed shell systems represents the ionisation potential from that level[15]), and this ionisation potential is, using the technique of photoelectron spectroscopy (93) measurable in principle and often, already in practice. The total energy can also be calculated, and represents the energy of formation of the system from infinitely separated nuclei (or nuclei with cores) and electrons. Net orbital populations, bond populations and gross populations are readily defined.

This treatment contains several sources of error, and an area of subjectivity, but these can be stated quite explicitly. The errors are due to relativistic effects (which will be relatively unimportant for outer 'valence' electrons), correlation, the assumed invariance of core orbitals, approximations made in evaluating or in ignoring integrals, and the restriction to a single configuration. Subjectivity is shown in the choice of basis set orbitals, and in the recipe for dividing bond population between atoms. The atomic orbitals chosen must represent a 'balanced set', if a realistic charge distribution is to be obtained and there may be disagreement about when a set can be taken as 'balanced'. The form of a given orbital (especially a transition metal d-orbital) is itself sensitive to the charge distribution. Finally, the criterion of goodness used (total energy) is much more sensitive to errors near the nuclei than in the region between them which is responsible for the bonding, and where the electron density is actually smaller.

The Wolfsberg-Helmholz method (94) is essentially a semi-empirical extended Hückel treatment. This involves the solution of a secular equation of form

$$|H_{ij} - S_{ij}E| = 0.$$

The diagonal elements of the Hamiltonian, H_{ii} or α_i, represent the energy of an atomic orbital in the field of the molecule. The off-diagonal elements, H_{ij}, are resonance terms between atomic orbitals i and j, and are taken as proportional to $FS_{ij}(\alpha_i + \alpha_j)/2$, or to $FS_{ij}\alpha_i\alpha_j$. The factor F may be assigned a value, or may be treated as a parameter. The eigenvectors of the secular determinant can be calculated, giving population analyses. The α values are taken from experimental ionisation potential data for the free atoms or ligands. These depend on charge, which in turn depends on the population analyses, as do the orbital sizes and overlaps. Some measure of self-consistency can be introduced by iteration, but this is not always done.

---

[15]) This interpretation does not allow for the reorganisation of the remaining electrons in the ion produced. The true ionisation potentials will therefore be lower than the best possible calculated values.

The procedure has the merit of simplicity, but few others. An atom in a molecule is surrounded by outside charges, and these will drastically affect the value of the α's. This 'Madelung-type energy', which is ignored here, has been shown (95) to make a major contribution to the binding energy. The recipe for evaluating the off-diagonal elements is highly arbitrary, and not even transformation-invariant under the operations of the molecular point group (96). The correct values of α, if the population analysis is to be meaningful, should correspond to marginal, not integral, ionisation potentials. Finally, the meaning of the ε values obtained is not clear.

III.2. Hartree-Fock Calculations

In a calculation on nickel tetracarbonyl (97) the Ni $1s$–$3p$, O $1s$ and C $1s$ were taken as 'core orbitals', and the Ni $3d$, $4s$, $4p$, O $2s$, $2p$ and C $2s$, $2p$ as the basis set. The nickel orbitals used were those of *Richardson, Nieuwpoort, Powell* and *Edgell* (98, 99) while the C and O orbitals were those given by *Ransil* (4) for these atoms in CO. The calculation was extended, with some loss of rigour, to $Co(CO)_4^-$ and $Fe(CO)_4^{2-}$, by allowing for the reduced charge of the metal nucleus without altering the orbitals in the core or in the basis set. The reviewer feels that this approximation exaggerates the trends along the series to an unknown degree. Ideally, the calculation should allow for expansion of the metal core orbitals as the nuclear attraction is weakened. The falling nuclear charge would then be, to some extent, compensated by less complete screening. The calculation as it stands gives positive energies for the highest occupied levels of the anions, a result which, if taken seriously, would imply that the ions should spontaneously dissociate into radicals and free electrons[16]. The calculation will magnify any trends in the metal-ligand π-bonding to an unknown degree; however, the estimate of these trends is admittedly (100) not quantitative.

Despite this defect, the results of the calculation are generally in accord with chemical intuition, and calculated values of $-\varepsilon$ for $Ni(CO)_4$ do not deviate too badly from observed ionisation potentials as shown by Table 1. One can imagine $Ni(CO)_4$ as being formed by the interaction of an excited ($3d^{10}$) nickel atom with four CO molecules. There is little resultant change in the CO σ-bond (3σ of CO), but there is appreciable mixing of the antibonding 4σ and 5σ, so as to give the highest ligand orbital more 'carbon lone pair' character. The net $d\sigma$ overlap population is small, justifying the interpretation of the 18-electron rule given in

---

[16] Assuming that the solvation or lattice stabilisation of an electron would not be much less than that of the anion.

section II(4). 1π, the CO π-bonding orbital, mixes considerably with 5σ but not with $3d$, while the vacant 2π ($\pi*$) overlaps 1π and $3d$, but has little interaction with 5σ. The assumptions of the conventional approach are thus upheld, while those of the 'short-circuiting theory', which would require 5σ to donate to 2π, are undermined. The net coefficient of the overlap between 1π and 2π is small and negative, both overall and in the non-bonding T_1 molecular orbitals. This suggests that the π-electron density becomes polarised away from the metal, rather than towards it,

Table 1. *Orbital Energies in Tetrahedral Carbonyls*[a])

Orbital	Main Components	Vertical I.P.[b]) (Ni(CO)$_4$)	Calculated ε [c])		
			Ni(CO)$_4$	Co(CO)$_4^-$	Fe(CO)$_4^{2-}$
$3e$	$3d, 2\pi$	empty	9.8	20.1	25.8
$6t_2$	$3d, 2\pi$	empty	7.1	13.1	20.1
$2t_1$	2π	empty	7.1	12.8	18.5
$2e$	$3d, 2\pi$	—9.7	—7.89	0.54	5.99
$5t_2$	$3d$	—8.8	—10.61	—1.36	7.89
$1t_1$	1π	—14.0	—16.60	—10.34	—3.81
$1e$	1π	{—14.8}	—18.23	—11.97	—5.99
$4t_2$	$1\pi, 5\sigma, 3d$		—18.23	—11.97	—5.99
$3t_2$	$4\sigma, 5\sigma$ ⎫	—17.2	—19.86	—14.15	—7.89
$3a_1$	4σ ⎭		—21.22	—15.24	—8.44
$2t_2$	$5\sigma, 4\sigma, 4p$		—26.94	—21.22	—15.51
$2a_1$	$5\sigma, 4s$		—31.56	—26.39	—22.04
$1a_1$	3σ		—53.06	—46.26	—38.91
$1t_2$	3σ		—53.60	—46.80	—39.73

[a]) Energies in e.v.
[b]) W. C. Price, private communication. Assignment (except for $2e$, $5t_2$) arbitrary.
[c]) Reference 97.

but that the effect is minimal. The CO π-bond is weakened by population of 2π from $3d$, as in the conventional picture, and also, to a smaller extent, by depopulation of 1π. The $3d$-π overlap takes place mainly in the e system. In the anions, back-bonding is increased, and (especially in Fe(CO)$_4^{2-}$) makes use of the $3d(t_2)$ electrons, though as stated above the results for the anions should be treated with caution.

The rule of orbital use is evidently satisfied. The rule of electron use is satisfied for all outermost electrons except the $3d(t_2)$ of Ni(CO)$_4$. This might lead us to expect that Ni(CO)$_4$ would be a nucleophile. In fact, it is not $Ni(CO)_4$ but $Co(CO)_4^-$ and $Fe(CO)_4^{2-}$, in which these orbitals do have a bonding function, that act as bases. The same loose central field that allows the $3d(t_2)$ orbitals to spread out over the carbonyls will allow them to be used by other ligands. Chemical stability, here as always, depends not on formal bonding character but on total energetics.

The problem of the metal hexacarbonyls is notionally an easier one, since $d(\sigma)$ and $d(\pi)$ are separated by symmetry, but presents a more formidable task because of the greater number of electrons and orbitals involved. This task has been performed (10), using the SCF—LCAO method, for the 'd^6' series $V(CO)_6^-$, $Cr(CO)_6$, $Mn(CO)_6^+$. The calculation resembles that on the tetracarbonyls in treating only M[Ar], C(1s) and O (1s) as core orbitals, and in the choice of the type of d-orbital (those for $V(-1)$ being found by extrapolation), but differs from it in the inclusion of $4d$ in the basis set, in treating each member of the series as a fresh problem, and in using separate, realistic values of the MC distance for each species. As before, it is found that the CO σ-bond is virtually unaltered, but that there is strong mixing of 4σ and 5σ. Metal-ligand π-bonding is limited by symmetry to the t_{2g} levels. As expected, the degree of π-bonding is greatest in $V(CO)_6^-$ (where the outer t_{2g} electron density is concentrated slightly more on 2π than on $3d$) and least in $Mn(CO)_6^+$. The *total* amount of back-donation (as measured by the depopulation of the $d(\pi)$ orbitals) varies from 2.0 for $Mn(CO)_6^+$ through 2.6 for $Cr(CO)_6$ to 3.1 for $V(CO)_6^-$. The corresponding figures for the tetracarbonyls are 1.7 ($Ni(CO)_4$), 3.4 ($Co(CO)_4^-$), 5.3 ($Fe(CO)_4^{2-}$); these last two figures are probably an overestimate). The amount of forward-donation (as measured by the gross population of $d(\sigma)$, $4s$, $4p$) is highest for $V(CO)_6^-$, a more surprising result. The total amounts are highest for the hexacarbonyls, but the amounts per CO group are very similar for the two series.

It does not seem possible at present to provide an 'explanation' in terms of bonding for the fact (27) that the metal-carbon bond energy is greater in $Cr(CO)_6$ than in $Ni(CO)_4$ (the average M—CO bond strength being estimated at 57.7 kcal/Mole, as against 45.7, with respect to normal CO and 'valence state' d^n metal). The bond energies involved are the small differences between the total energies for the separated and united fragments, which are the calculated quantities. It is tempting to attribute the existence of cationic hexacarbonyls, while cationic tetracarbonyls are at present unknown, to superior σ-bonding. It should be remembered, however, that existence in chemistry is essentially competitive. $Mn(CO)_6^+$ $AlCl_4^-$ is stable *with respect to* $Mn(CO)_5Cl + AlCl_3 + CO$. The failure to prepare $Cu(CO)_4^+$ may simply be due to the smaller radius and lower-lying $4s$, p orbitals of Cu^+, and the correspondingly increased stability of Cu(I) complexes of more basic ligands, rather than to any weakness in the hypothetical Cu^+—CO bonds.

III.3. Wolfsberg-Helmholz and Related Calculations

Approximate calculations on the metal carbonyls will not (except for involatile species) give us better ionisation potentials than those obtainable by photoelectron spectroscopy, nor will they give us more insight into the nature of the bonding than the SCF calculations, where these are available. Despite this, approximate calculations remain of value, for several reasons. It is of inherent interest to know just how bad the approximate methods are, and whether modifications of the calculation procedure make them much better. Complicated systems can only be dealt with by the approximate methods, and it is important to know, from studies of simpler models, how much confidence we are to have in the results. Finally, the approximate methods can be applied to *series* of compounds (without prohibitively extravagant use of computer time); we may then hope to relate trends in calculated quantities both to their underlying causes and to their observable consequences (as when a calculated change in overlap population is explained in terms of changing electronegativity, and related to a change in stretching frequency).

The metal hexacarbonyls have been investigated using both the simple *Wolfsberg-Helmholz* method, and several refinements designed to meet in part the criticisms (III—9) to which this open. Some results of these calculations are compared for $Cr(CO)_6$ (Table 2) with those of the more exact calculation (*10*) and with experimental data.

Table 2. *Calculated and Experimental Orbital Energies in* $Cr(CO)_6$

Orbital	Principal Compounds	Calculated energy (eV)						Experimental Vertical I.P.[c]
		$(10)^{a}$	$(9)^{b}$	$(8b)^{b}$	$(11)^{b}$	$(12)^{b}$	$(13)^{b}$	
t_{2u}	π^*		—7.11	—3.67		—10.1		
t_{1u}	π^*		—7.88	—3.89		—9.8		
e_g	d		—3.19	—4.08	+3.67	—4.8		
t_{2g}	d, π^*	—8.19	—10.02	—8.37	—6.46	—12.8	—8.24	8.4
t_{1u}	σ, π	—14.02	—13.27	—12.87	—6.74	—14.3	—14.01	13.32
t_{1g}	π	—14.90	—13.10	—15.62	—5.92	—15.4	—16.62	
t_{2u}	π	—14.95	—13.10	—16.43	—7.54	—15.6	—16.64	14.12
t_{1u}	σ, π	—14.97	—14.13	—17.02	—12.79	>—15.6	—16.60	
a_{1g}	σ	—15.23	—14.16	—14.97	—14.58	—15.1	—13.61	14.49
e_g	σ	—16.14	—14.40	—13.14	—13.77	—14.5	—14.06	15.2
t_{2g}	π	—16.24	—13.30	—18.51	—12.01	>—15.6	—16.71	15.6

[a]) A priori SCF.
[b]) Semi-empirical
[c]) Data of *D. W. Turner, C. Baker*, and *W. C. Price*, quoted by 8(*b*), 10, 13. Assignments not experimentally accessible.

In the simplest *Wolfsberg-Helmholz* calculation (*9*) the oxygen lone pairs, CO σ-bonding electrons, and the [Ar] configuration of Cr (O) were taken as part of the core. The $3d$, $4s$, p of Cr, $2p(\pi)$ of C and O and the carbon lone pair (represented by $2p(\sigma)$; this procedure will tend to exaggerate carbon to metal σ-bonding) were taken as the basis set. The orbital exponents were derived from Slater's rules, which would be expected to exaggerate the metal-ligand overlap, and there was no iteration to self-consistency.

The more recent papers (*8b, 11—13*) have improved on this simple *Wolfsberg-Helmholz* procedure in several ways. All of them have iterated their calculations to self-consistency, using the procedure of *Gray et al.* (*101*) to allow for the effect of charge on the ionisation potential of the metal. One calculation (*12*) applies a Madelung correction (*95*) to the on-diagonal energy terms of the secular determinant but not to the off-diagonal terms (an omission that can be criticised (*102*), but it is stated that inclusion of the such off-diagonal corrections leads to unacceptable results). Two calculations use the *Wolfsberg-Helmholz* form $FS_{ij}(\alpha_i + \alpha_j)/2$ for the off-diagonal elements; in one case (*12*) F is fixed as 1.75, while in the other (*8b*) it is treated as a parameter. The others (*11, 13*) employ the logically sounder Cusachs approximation (*103*) $(2 - [S_{ij}]) \, (\alpha_i + \alpha_j) S_{ij}/2$, which is invariant under the operations of the molecular point-group (*96*), for the off-diagonal elements of the Hamiltonian. The metal orbitals are either self-consistent, or analytical approximations to self-consistent orbitals, while a variety of recipes are used to find the ligand orbitals. In two cases (*12, 13*) the procedure for choosing an energy for the CO π^* orbital depends on the use of spectroscopic data, implicit in which is the illegitimate (*102, 104, 105*) (see also the discussion of ultraviolet spectra in Part B of this review) equation of excitation energies with orbital energy differences, while in a third (*8b*) it is treated as a parameter, and adjusted (quite as questionably) to relate the $d-d$ splitting to the calculated one-electron energy difference between $t_{2g}(\pi)$ and $e_g(\sigma^*)$ in the complex.

In each case, the calculation is extended to a series of compounds; $V(CO)_6^-$, $Cr(CO)_6$, $Mo(CO)_6$, $W(CO)_6$, $Mn(CO)_6^+$, $Re(CO)_6^+$ (*8b*) (the method has also been applied by the same school to $Fe(CO)_5$ (*106*)) $Cr(CO)_6$, $C_6H_6Cr(CO)_3$, $C_6H_5NH_2Cr(CO)_3$ (*11*); $Cr(CO)_6$, $Fe(CO)_5$, $Ni(CO)_4$ (*12*); $Cr(CO)_6$, $Mo(CO)_6$, $W(CO)_6$ (*13*); the approximate nature of the calculations is freely admitted and it is stated that part of their purpose is the study of trends. This is a realistic goal provided that the variation in computational error along the series considered is not so great as to distort the trend of interest (an assumption that is difficult to validate without using the very data to be explained). Success achieved in the discussion of trends does not and cannot constitute a justification of the *absolute* values calculated. Such a justification could only come (in the absence of any sound theoretical basis for semi-empirical calculations in general) from a degree of agreement between calculation and experiment, and between one calculation and another, that is so far lacking.

Note added to proof

The crystal structure of $Rh_2[P(OPh)_3]_2(COD)Cl_2$ shows puckering of the dimeric units, although these are not stacked and bonding between units is impossible; thus criticism of (II(8)) of the 'bend bond' description of $(Rh(CO)_2Cl)_2$ (*49*) receives support (*107*). The 86-electron rule for octahedral metal carbonyl clusters (II(9)) is further and most convincingly exemplified by $H_2Ru_6(CO)_{18}$ (*108*). A *priori* calculations for $Mn(CO)_5Br$ (*109*) show impressive agreement with the photoelectron spectrum (*93*), correctly predicting that the highest occupied orbitals are π-donating $p(x, y)$ of Br. Semi-empirical calculations can be greatly improved by a procedure that includes both diagonal and non-diagonal ligand field terms; the procedure has been applied to $Cr(CO)_6$, $Fe(CO)_5$, $Fe(CO)_4^{2-}$, $Co(CO)_4^{-}$ and $Ni(CO)_4$ (*110*), and there is a satisfying measure of agreement both with the a *priori* calculation and with the available experimental data.

It is a pleasure to thank Mrs. *R. M. Canadine*, Professor *P. Chini*, Professor *R. F. Fenske*, Dr. *W. C. Nieuwpoort*, Professor *W. C. Price* and Dr. *D. W. Turner* for stimulating correspondence, and in some cases for access to unpublished results.

References

1. See e.g. (a) *Chatt, J., Pauson, P. L., Venanzi, L. M.*: in Organometallic Chemistry, ed. *Zeiss, H.*, Reinhold, N. Y., 1960, p. 468 (general); b) *Abel, E. W.*: Quart. Rev. Chem. Soc. (London) *17*, 133 (1963) (general); c) *Wender, I., Pino, P.*: (ed.) Organic Synthese via Metal Carbonyls, *1*, (1968), Interscience 1968 (general and applications); d) *Dobson, G. R., Stotz, I. W., Sheline, R. K.*: Advan. Inorg. Chem. Radiochem. *8*, 1 (1966) (Group VI carbonyls); e) *Abel, E. W., Stone, F. G. A.*: Quart. Rev. Chem. Soc. (London) *23*, 325 (1969) (structural considerations); f) *King, R. B.*: Advan. Organomet. Chem. *2*, 157 (1964) (carbonyl anions); g) *Chini, P.*: Inorg. Chim. Acta Reviews *2*, 31 (1968) (polynuclear carbonyls); h) *Brown, D. A.*: Inorg. Chim. Acta Reviews *1*, 35 (1967), *Strohmeier, W.*: Fortschr. Chem. Forsch. *10*, 306 (1968), *Werner, H.*: Angew. Chem. Intern. Ed. *7*, 930 (1968) (substitution reactions); i) *Hieber, W.*: Advan. Organometal. Chem. *8*, 1 (1970) (personal survey); j) *Abel, E. W., Tyfield, S. P.*: Advan. Organomet. Chem. *8*, 117 (1970) (carbonyl cations).
2. a) *Pauling, L.*: Nature of the Chemical Bond, Cornell U. P., 1939; b) *Gilliland, W. L., Blanchard, A. A.*: J. Amer. Chem. Soc. *48*, 410 (1926).

3. a) *Graham, W. A. G., Stone, F. G. A.:* J. Inorg. Nuclear Chem. *3,* 164 (1956);
   b) *Armstrong, D. R., Perkins, P. G.:* J. Chem. Soc. (A), 1044 (1969).
4. *Ransil, B. J.:* Rev. Mod. Phys. *32,* 245 (1960).
5. *Jaffe, H. H., Doak, G. O.:* J. Chem. Phys. *21,* 196 (1953).
6. See e.g. *Orgel, L. E.:* Introduction to Transition Metal Chemistry, Methuen and Co. 1960.
7. *Bethe, H. A.:* Ann. Phys. Lpz. *3,* 133 (1929).
8. a) *Gray, H. B., Beach, N. A.:* J. Amer. Chem. Soc. *85,* 2922 (1963); b) *Beach, N. A., Gray, H. B.:* J. Amer. Chem. Soc. *90,* 5713 (1968).
9. *Bdn, M. I., Fényi, Sz., Hegyhati, M.:* Theory and Structure of Complex Compounds (Wroclaw Conference, 1962), ed. *Jezowska-Trzebiatowska, B.:* Pergamon Press, 1964, p. 195.
10. *Caulton, K. G., Fenske, R. F.:* Inorg. Chem. *7,* 1273 (1968).
11. *Carroll, D. G., McGlynn, S. P.:* Inorg. Chem. *7,* 1285 (1968).
12. *Schreiner, A. F., Brown, T. L.:* J. Amer. Chem. Soc. *90,* 3366, 5947 (1968).
13. *Brown, D. A., Rawlinson, Sr. R. M.:* J. Chem. Soc. (A), 1530 (1969).
14. *Brockway, L. O., Ewens, R. V. G., Lister, M. W.:* Trans. Farad. Soc. *34,* 1350 (1938).
15. *Shufler, S. L., Sternberg, H. W., Friedel, R. A.:* J. Amer. Chem. Soc. *78,* 2687 (1956).
16. *Sidgwick, N. V.:* Electronic Theory of Valency, O. U. P., 1927, p. 163.
17. *King, R. B.:* Advan. Chem. Ser. *62,* 203 (1967).
18. *Blanchard, A. A.:* Chem. Revs. *21,* 7 (1937).
19. See e.g. *Gillespie, R. J., Nyholm, R. S.:* Quart. Revs. Chem. Soc. (London) *11,* 339 (1957).
20. *Rettig, M. F., Drago, R. S.:* J. Amer. Chem. Soc. *91,* 3432 (1969).
21. *McBride, D. W., Pruett, R. L., Pitcher, E., Stone, F. G. A.:* J. Amer. Chem. Soc. *84,* 497 (1962).
22. *Kettle, S. F. A.:* J. Chem. Soc. (A) 1013 (1966).
23. *Person, W. B., Williams, D. G.:* J. Phys. Chem. *61,* 1017 (1957).
24. *Bergel, F.:* Chem. Ber. *62B* 490 (1929).
25. *Van Krevelen, D. W., Baans, C. M. E.:* J. Phys. Colloid Chem. *54,* 370 (1950).
26. *Edgell, W. F., Huff, J., Thomas, J., Lehman, H., Angell, C., Asato, G.:* J. Amer. Chem. Soc. *82,* 1254 (1960).
27. *Skinner, H. A.:* Advan. Organometal. Chem. *2,* 110 (1964).
28. *Basolo, F., Wojcicki, A.:* J. Amer. Chem. Soc. *83,* 520 (1961).
29. *Kettle, S. F. A.:* Inorg. Chem. *4,* 1661 (1965).
30. — J. Chem. Soc. (A) 420 (1966).
31. — private communication.
32. *Cotton, F. A.:* Inorg. Chem. *3,* 702 (1964).
33. *Powell, H. M., Ewens, R. V. G.:* J. Chem. Soc. 286 (1939).
34. *Cotton, F. A.:* in Modern Coordination Chemistry, ed. *Lewis, J., Wilkins, R. G.,* Interscience, 1960.
35. *Bellamy, L. J.:* Infrared Spectra of Complex Molecules, Methuen/Wiley 1954, p. 114.
36. *Beck, W., Hieber, W., Tengler, H.:* Chem. Ber. *94,* 862 (1961).
37. *Breslow, D. S., Heck, R. F.:* Chem. and Ind. (London), 467 (1960).
38. *Mills, O. S.:* Acta Cryst. *11,* 620 (1958).
39. *Sumner, G. G., Klug, H. P., Alexander, L. E.:* Acta Cryst. *17,* 732 (1964).
40. *Braterman, P. S., Chini, P.:* Proc. Intl. Symp. Metal Carbonyls, Venice, 1968, B.7.
41. *Pople, J. A.:* Quart. Revs. Chem. Soc. (London), *11,* 273 (1957).

42. *Dahm, D. J., Jacobson, R. A.:* J. Amer. Chem. Soc. *90,* 5106 (1968).
43. *Snow, A. I., Rundle, R. E.:* Acta Cryst. *4,* 348 (1951).
44. *Lewis, P. H., Rundle, R. E.:* J. Chem. Phys. *16,* 552 (1948).
45. *Dahl, L. F., Wei, C.-H.:* Inorg. Chem. *2,* 328 (1963).
46. *Wei, C.-H., Dahl, L. F.:* Inorg. Chem. *4,* 1 (1965).
47. *Hayter, R. G.:* Inorg. Chem. *3,* 711 (1964); see also (56).
48. *Occam, William of.:* Quodlibeta i, Q. 3, Summa Logicae, pars 1, cap. 12.
49. *Dahl, L. F., Martell, C., Wampler, D. L.:* J. Amer. Chem. Soc. *83,* 1761 (1961).
50. *Braterman, P. S., Thompson, D. T.:* J. Organometal. Chem. *10,* P 11 (1967); J. Chem. Soc. (A) 1454 (1968).
51. *Mais, R. H. B., Owston, P. G., Thompson, D. T.:* J. Chem. Soc. (A) 1735 (1967).
52. *Thomas, J. T., Robertson, G. H., Cox, E. G.:* Acta Cryst. *11,* 599 (1958).
53. *Carty, P., Walker, A., Mathew, M., Palenik, G, J.:* Chem. Commun. 1314 (1969).
54. *Kaesz, H. D., Fellmann, W., Wilkes, G. R., Dahl, L. F.:* J. Am. Chem. Soc. *87,* 2753 (1965).
55. *Doedens, R. J., Robinson, W. T., Ibers, J. A.:* J. Am. Chem. Soc. *89,* 4323 (1967).
56. *Dahl, L. F., Rundle, R. E.:* Acta Cryst. *16,* 419 (1963).
57. *Cotton, F. A.:* Quart. Revs. Chem. Soc. (London), *20,* 389 (1966) and references therein.
58. *Corey, E. R., Dahl, F. L., Beck, W.:* J. Am. Chem. Soc. *85,* 1202 (1963).
59. *Wei, C.-H., Dahl, L. F.:* J. Am. Chem. Soc. *91,* 1351 (1969).
60. *Huggins, D. K., Fellman, W., Smith, J. M., Kaesz, H. D.:* J. Am. Chem. Soc. *86,* 4841 (1963).
61. *Bennett, M. J., Cotton, F. A., Winquist, B. H. C.:* J. Am. Chem. Soc. *89,* 5366 (1967).
62. *Corey, E. R., Dahl, L. F.:* Inorg. Chem. *1,* 521 (1962).
63. *Wilkes, G. R.:* Ph. D. Thesis, University of Winsconsin, Madison, 1965, quoted in (72).
64. *Hock, A. A., Mills, O. S.:* Advances in the Chemistry of the Coordination Compounds, ed. *Kirschner, S.:* New York: Macmillan 1961, p. 640.
65. *Braterman, P. S.:* Chem. Commun. 91 (1968).
66. *Newman, M. A., King, R. B., Dahl, L. F.:* quoted in (61).
67. *Braterman, P. S.:* J. Chem. Soc. (A), 2907 (1968).
68. *Chini, P.:* Chem. Commun. 440 (1967).
69. *Bau, R., Fontal, B., Kaesz, H. D., Churchill, M. R.:* J. Am. Chem. Soc. *89,* 6374 (1967).
70. *Vranka, R. G., Dahl, L. F., Chini, P., Chatt, J.:* J. Am. Chem. Soc. *91,* 1574 (1969).
71. *Doedens, R. J., Dahl, L. F.:* J. Am. Chem. Soc. *88,* 4847 (1966).
72. *Wei, C.-H., Dahl, L. F.:* J. Am. Chem. Soc. *88,* 1821 (1966).
73. *Wei, C.-H., Dahl, L. F:* Inorg. Chem. *6,* 1229 (1967).
74. *Chini, P., Ercoli, R.:* Gazz. Chim. Ital. *88,* 1170 (1958).
75. *Sutton, P. W., Dahl, L. F.:* J. Am. Chem. Soc. *89,* 261 (1967).
76. *Vahrenkamp, H., Uchtman, V. A., Dahl, L. F.:* J. Am. Chem. Soc. *90,* 3272 (1968).
77. *Dahl, L. F.:* Proc. 3rd Intl. Symp. Organometal. Chem., Munich, (1967).
78. *Guggenberger, L. J.:* Chem. Commun. 512 (1968).
79. *Albano, V., Chini, P., Scatturin, V.:* Chem. Commun. 163 (1968).
80. *Chini, P., Albano, V.:* J. Organometal. Chem. *15,* 433 (1968).
81. *Albano, V., Bellon, P. L., Chini, P., Scatturin, V.:* J. Organometal. Chem. *16,* 461 (1969).

82. *Johnson, B. F. G., Johnston, R. D., Lewis, J.:* Chem. Commun. 1057 (1967); *Mason, R., Robinson, W.:* Chem. Commun. 468 (1968).
83. *Braye, E. H., Dahl, L. F., Hubel, W., Wampler, D. L.:* J. Am. Chem. Soc. *84*, 4633 (1962).
84. *Albano, V. G., Bellon, P. L.:* J. Organometal. Chem. *19*, 403 (1969).
85. —, —, *Ciani, G. F.:* Chem. Commun. 1024 (1969).
86. *Sirigu, A., Bianchi, M., Benedetti, E.:* Chem. Commun. 596 (1969).
87. *Cotton, F. A., Haas, T. E.:* Inorg. Chem. *3*, 10 (1964).
88. *Kettle, S. F. A.:* J. Chem. Soc. (A) 314 (1967).
89. *Braterman, P. S.:* unpublished results.
90. *Slater, J. C.:* Quantum Theory of Atomic Structure, New York: McGraw-Hill, 1960, Chapters 9, 15, 16, 17.
91. *Moskowitz, J. W.:* J. Chem. Phys. *43*, 60 (1965).
92. *Roothaan, C. C. J.:* Rev. Mod. Phys. *23*, 69 (1951).
93. See e.g. *Evans, S., Green, J. C., Green, M. L. H., Orchard, A. F., Turner, D. W.:* Disc. Farad. Soc. *47*, 112 (1969).
94. *Wolfsberg, M., Helmholz, L.:* J. Chem. Phys. *20*, 837 (1952).
95. *Jørgensen, C. K.:* Structure and Bonding *1*, 3 (1966).
96. *Carroll, D. G., McGlynn, S. P.:* J. Chem. Phys. *45*, 3827 (1966).
97. *Nieuwpoort, W. C.:* Philips Res. Rept. *20*, Suppl. 6 (1965).
98. *Richardson, J. W., Nieuwpoort, W. C., Powell, R. R., Edgell, W. F.:* J. Chem. Phys. *36*, 1057 (1962).
99. *Richardson, J. W., Powell, R. R., Nieuwpoort, W. C.:* J. Chem. Phys. *38*, 796 (1963).
100. *Nieuwpoort, W. C.:* private communication.
101. *Basch, H., Viste, A., Gray, H. B.:* J. Chem. Phys. *44*, 10 (1966).
102. *Canadine, R. M., Hillier, I. H.:* J. Chem. Phys. *50*, 2989 (1969).
103. *Cusachs, L. C.:* J. Chem. Phys. *43*, 1575 (1965).
104. *Canadine, R. M.:* private communication.
105. *Fenske, R. F.:* private communication.
106. *Dartiguenave, Y., Dartiguenave, M., Gray, H. B.:* Bull. Chem. Soc. France 4223 (1969).
107. *Coetzer, J., Gafner, G.:* Acta Cryst. B *26*, 985 (1970).
108. *Churchill, M. R., Wormald, J., Knight, J., Mays, M. J.:* Chem. Commun. 458 (1970).
109. *Fenske, R. F., DeKock, R. L.:* Inorg. Chem. *9*, 1053 (1970).
110. *Hillier, I. H.:* J. Chem. Phys. *52*, 1948 (1970).

Received April 28, 1971

On the General Theory of Magnetic Susceptibilities of Polynuclear Transition-metal Compounds

Systems with Two or Three Spins

J. S. Griffith *

Chemistry Department, Indiana University, Bloomington, Indiana, USA

Table of Contents

1. Introduction

Polynuclear transition-metal compounds are currently of considerable interest both in inorganic chemistry (for reviews see Ref. (*1*) and (*2*)) and in biochemistry. In the latter field attention has mainly concentrated on iron-sulphur proteins (*3, 4*) although other examples are cytochrome oxidase (*5, 6, 7*) the copper-containing ceruloplasmin (*8*) and respiratory carrier haemocyanin (*9*), the iron and molybdenum-containing nitrogen fixing enzyme (*10*) and probably the oxygen-evolving photosynthetic system which may contain a group of manganese ions bound in a metallo-protein (*11, 12*).

---

* Now at: Basel Institute for Immunology, 487 Grenzacherstraße, CH 4058, Basel, Switzerland

The number of different possible polynuclear compounds of even moderate complexity is extremely large and in order to have detailed understanding of their magnetic properties it would be necessary to discuss each such compound separately and in detail. It is the purpose of this paper, however, to show that there is quite a body of general theory which may be given and which goes far towards giving a general solution to some of the theoretical problems arising from such compounds.

We shall concern ourselves with systems, such as the nickel acetylacetonate trimer (13) or the copper acetate dimer (14, 15), in which the metal ions are well separated and the interaction energy small. The important question of the mechanism of this interaction will not be discussed here (16, 17, 18). We then address ourselves to the following question: given a polynuclear system containing n ions, each of whose properties we know separately, and given a certain form of interaction between each pair of ions, how do we calculate the magnetic properties of the coupled system in terms of the parameters describing the constituent ions and the interaction. We are only interested in the thermally occupied states and thus contemplate writing a Hamiltonian for the whole spin system in the absence of an external magnetic field as:

$$\mathcal{H} = \sum_{i=1}^{n} \mathcal{H}_i + \sum_{i<j}^{n} \mathcal{V}_{ij} \tag{1}$$

where $\mathcal{H}_i$ refers to the ith ion alone and the interaction $\mathcal{V}_{ij}$ is often taken in the form $J_{ij}\mathbf{S}_i \cdot \mathbf{S}_j$. The formulation in Eq. (1) contains the explicit assumption that we may neglect "three-body" or higher interactions, e.g. terms of the kind $K_{hij}\mathbf{S}_h \cdot (\mathbf{S}_i \wedge \mathbf{S}_j)$, although we shall consider these briefly in section 3E. It also has the implicit assumption that states involving charge transfer between the ions are unimportant (15).

From the point of view of theoretical difficulty there is a great difference between ions, such as octahedrally-coordinated Ni^{2+} or Cr^{3+}, which have only spin degeneracy in their thermally-occupied states and those, such as octahedral high spin Fe^{2+} or low spin Fe^{3+}, which also have spatial degeneracy. Because we may regard this spatial degeneracy as coming from a pseudo-angular momentum vector (20), with $L=1$, it follows that the total number of "angular momenta" connected with a polynuclear system is really $N = n_1 + 2n_2$ where n_1 is the number of ions with only spin degeneracy and n_2 the number with both spin and spatial degeneracy. The complexity of the corresponding theoretical problem depends primarily on this number N rather than on n. Thus, in general, the problem of two of the second kind of ion presents a difficulty similar to and comparable with that of four of the first kind. In this paper we shall restrict ourselves to polynuclear systems having two or three constituent ions, each having only spin degeneracy. It is hoped to discuss more complicated systems in a later paper.

Finally, we shall be considering mainly magnetic susceptibilities rather than paramagnetic resonance. There is a good reason for this. In paramagnetic resonance we need to include terms in the Hamiltonian which give effects comparable with or smaller than the magnetic field energies (typically about 0.3 cm^{-1}). This means that we must include terms, such as the magnetic dipole-dipole interaction, which do not usually commute with the total spin S of the system. However because of the statistical averaging involved in a magnetic susceptibility, energies small compared with kT may be largely neglected (*19, 20*). Consequently, except at very low temperatures, we are generally only concerned with interactions which derive ultimately from the electrostatic interaction between the electrons. This means that $\Sigma \mathscr{V}_{ij}$ commutes with the total spin S for the whole system, while each $\mathscr{V}_{ij}$ separately commutes with the total spin $S_i + S_j$ for the two-ion system $i + j$. Except in Sect. (4) we shall make this assumption. In particular, this condition is satisfied by the conventional Heisenberg interaction $\mathscr{V}_{ij} = J_{ij} S_i \cdot S_j$, but as we shall see later it is also satisfied by more general interactions. Except in Sect. (4), we assume also that $\mathscr{H}_i$ commutes with S_i, *i.e.* that the $(2 S_i + 1)$ occupied states of the constituent ion i would be degenerate in the absence of the interaction. In this case the terms $\mathscr{H}_i$ merely give an additive constant to all energies and this may be conveniently taken equal to zero and the $\mathscr{H}_i$ dropped from Eq. (1).

2. Two-spin Systems

A. Energies

We now suppose we have two ions with respective spins S_1, S_2 and no spatial degeneracy. The total spin for the system is

$$S = S_1 + S_2 \tag{2}$$

and a form of interaction[1]) which has often been assumed (*1, 2, 21*) is

$$\mathscr{V} = J S_1 \cdot S_2 \tag{3}$$

We shall use this interaction in the present section although in Sect. 3 it will emerge that a more general interaction must often be considered. Because of the relation

$$2 S_1 \cdot S_2 = S^2 - S_1^2 - S_2^2 \tag{4}$$

---

[1]) Chemists often write $-J$ or $-2J$ in place of J in Eq. (3), but the simpler form seems preferable.

it follows that the energies of the coupled system are given by

$$E(S) = \tfrac{1}{2} J \{S(S+1) - S_1(S_1+1) - S_2(S_2+1)\} \tag{5}$$

Thus with this form for $\mathscr{V}$ the energies satisfy a Landé interval rule:

$$E(S) - E(S-1) = JS \tag{6}$$

B. g Tensor and Nuclear Hyperfine Interaction

Before continuing, it is interesting to note that, from an abstract point of view, the calculation of the g-tensor and magnetic susceptibility for our present coupled system presents a closely similar problem to that which appears in two physically very different situations. The first occurs in the rare earths, as treated by *Van Vleck* and *Frank* (19), where there are two independent angular momenta S and L for the system and instead of (2) and (3) we have

$$\boldsymbol{J} = \boldsymbol{S} + \boldsymbol{L}, \tag{2'}$$

$$\mathscr{V} = \lambda \boldsymbol{S} \cdot \boldsymbol{L} \tag{3'}$$

The second case arises with T_1 and T_2 terms for transition-metal ions in octahedral or tetrahedral symmetry when we work with a pseudo-angular momentum $\boldsymbol{L}$, with $L=1$, to describe the orbital degeneracy (see Ref. (20), Sects. (5.6) and (10.2)) and again have Eqs. like (2') and (3'). In the notation of Sect. 1, these are the case $n = n_2 = 1$, $N = 2$.

If it is correct to write the interaction between an external magnetic field $\boldsymbol{H}$ and our two ions in the isotropic form

$$\mathscr{H} = \beta g_1 \boldsymbol{H} \cdot \boldsymbol{S}_1 + \beta g_2 \boldsymbol{H} \cdot \boldsymbol{S}_2 \tag{7}$$

then the expression for the g value for the coupled system is (22):

$$g = \tfrac{1}{2}(g_1 + g_2) + \tfrac{1}{2}(g_1 - g_2) \cdot \frac{S_1(S_1+1) - S_2(S_2+1)}{S(S+1)} \tag{8}$$

which is closely related both to the formula holding for the transition-metal ion case mentioned above (Ref. (20), Eq. (5.59)) and to that for the Landé g-factor for an atom in Russell-Saunders coupling, which is

$$g = \tfrac{3}{2} + \frac{S(S+1) - L(L+1)}{2J(J+1)} \tag{9}$$

and which was used in *Van Vleck* and *Frank's* calculation.

However it is not necessarily true that the g-tensors for the individual ions can be written in an isotropic form. If not, then provided the principal axes are parallel for the two ions it is evident, by choosing the axis of interest to be the Z-axis, that Eq. (8) holds for each of these axes separately but generally with different g, g_1 and g_2 for each axis. We now ask what happens in the more general case that the two individual ions do not necessarily have parallel axes for the g-tensors. Then we have

$$\mathcal{H} = \beta \sum_{jk} (g_{jk}^{(1)} S_{1k} + g_{jk}^{(2)} S_{2k}) H_j \tag{10}$$

and have to find a spin-Hamiltonian

$$\mathcal{H}'(S) = \beta \sum_{jk} g_{jk} H_j S_k \tag{11}$$

which produces the same matrix elements as $\mathcal{H}$ within the $(2S + 1)$ states $|SM\rangle$ of the coupled system which have total spin S. First consider the parts involving the z components of the spins. These give a contribution to $\langle SM | \mathcal{H} | SM' \rangle$ of

$$\beta \sum_{j} \{g_{jz}^{(1)} H_j \langle SM | S_{1z} | SM' \rangle + g_{jz}^{(2)} H_j \langle SM | S_{2z} | SM' \rangle\}$$

and to $\langle SM | \mathcal{H}' | SM' \rangle$ of

$$\beta M \delta_{MM'} \sum_{j} g_{jz} H_j$$

The matrix elements of S_{1z} and S_{2z} in the coupled scheme are of course well known (20, 23) whence by comparing the coefficients of H_j we obtain g_{jz} in terms of $g_{jz}^{(1)}$ and $g_{jz}^{(2)}$. It is clear from symmetry that a relation of exactly the same form must also hold for g_{jx} and g_{jy} and hence we get the general formula

$$g_{jk} = \tfrac{1}{2}(g_{jk}^{(1)} + g_{jk}^{(2)}) + \tfrac{1}{2}(g_{jk}^{(1)} - g_{jk}^{(2)}) \cdot \frac{S_1(S_1 + 1) - S_2(S_2 + 1)}{S(S + 1)} \tag{12}$$

We can also obtain Eq. (12) for $k = x,y$ directly by considering the matrix elements of the S_{1x}, S_{1y}, etc., parts of $\mathcal{H}$ and $\mathcal{H}'$. Note that Eq. (12) has exactly the same form as Eq. (8) but it is much more general than it, because it makes no assumption either of isotropicity or about the relative angles of the principal axes of the g-tensors for the constituent ions.

Although we are not primarily interested in paramagnetic resonance here, it is worth pointing out that the nuclear hyperfine interaction may be treated in an exactly analogous manner. If the nuclear interaction Hamiltonian is

$$\mathcal{H} = \sum_{jk} (a_{jk}^{(1)} I_{1j} S_{1k} + a_{jk}^{(2)} I_{2j} S_{2k}) \tag{13}$$

91

then for the coupled system we get

$$\mathscr{H}' = \sum_{jk} (A_{jk}^{(1)} I_{1j} + A_{jk}^{(2)} I_{2j}) S_k \tag{14}$$

where

$$A_{jk}^{(1)} = \frac{S(S+1) + S_1(S_1+1) - S_2(S_2+1)}{2S(S+1)} a_{jk}^{(1)},$$

$$A_{jk}^{(2)} = \frac{S(S+1) - S_1(S_1+1) + S_2(S_2+1)}{2S(S+1)} a_{jk}^{(2)}. \tag{15}$$

C. Magnetic Susceptibility

Eq. (12) does not necessarily give us directly the principal g-values for the coupled system and in the general case these may have to be obtained computationally from the g-tensor. One might think, therefore, that the susceptibility of the coupled system could also only be obtained implicitly, except in special cases. However, because of the statistical averaging over states which occurs for the susceptibility, it turns out that we can give closed explicit formulae for χ, as we now see.

There are various contributions to the susceptibility and we shall discuss these in turn. The first comes from matrix elements of the magnetic field energy within the set of $(2S + 1)$ states having a given total spin S. Take the field H along a general direction, so that

$$\boldsymbol{H} = (h_x, h_y, h_z) H \tag{16}$$

with the h_i direction cosines for the vector $\boldsymbol{H}$. The interaction with the magnetic field is

$$\mathscr{H} = \beta H \sum_{jk} g_{jk} h_j S_k \tag{17}$$

where g_{jk} is given in Eq. (12). Now write V for the matrix of the operator $H^{-1}\mathscr{H}$. Its non-vanishing matrix elements are

$$V_{MM} = \beta M \sum_j g_{jz} h_j$$

$$V_{MM-1} = \tfrac{1}{2}\beta (S - M + 1)^{\frac{1}{2}} (S + M)^{\frac{1}{2}} \sum_j (g_{jx} - ig_{jy}) h_j \tag{18}$$

and the complex conjugates of the latter elements. Then

$$\chi = \frac{N}{kT(2S+1)} \sum E_i^2 \tag{19}$$

where the E_i are the $(2S + 1)$ eigenvalues of the matrix V.

As the squares E_i^2 are the eigenvalues of the matrix V^2, it follows that [2]).

$$\Sigma\, E_i^2 = \sum_{M,N} |V_{MN}|^2 = \sum_{M=-S}^{+S} V_{MM}^2 + 2 \sum_{M=-S+1}^{S} |V_{MM-1}|^2 \qquad (20)$$

We now introduce the values for the matrix elements of V, as given in Eq. (18), and perform the indicated summations over M. We obtain

$$\Sigma\, E_i^2 = (1/3)\, S\,(S+1)(2S+1)\, \beta^2 \sum_k (\sum_j g_{jk} h_k)^2 \qquad (21)$$

Substitute this in Eq. (19) to give

$$\mu^2 = \frac{3\,kT\chi}{N\beta^2} = S\,(S+1) \sum_k (\sum_j g_{jk} h_j)^2 \qquad (22)$$

This result refers to a particular field direction. If we now average over angle, using $\overline{(h_j h_l)} = (1/3)\, \delta_{jl}$, we obtain

$$\overline{\mu^2} = (1/3)\, S\,(S+1) \sum_{jk} g_{jk}^2 \qquad (23)$$

which is appropriate for a polycrystalline specimen. In case the g-tensor is isotropic, i.e. $g_{jk} = g\delta_{jk}$, this reduces to the usual extended spin-only formula

$$\overline{\mu^2} = g^2\, S\,(S+1) \qquad (24)$$

The next contribution to the susceptibility comes from matrix elements of the magnetic field energy between states differing in their S values. Because of the selection rule $\Delta S = 0, \pm 1$ for S_1 and S_2, it follows from Eq. (10) that the only non-vanishing matrix elements of this type are between states differing by one unit in S. Now rewrite Eq. (10) as

$$\mathscr{H} = \beta \sum_{jk} (g_{jk}^{(1)} S_k + (g_{jk}^{(2)} - g_{jk}^{(1)})\, S_{2k}\}\, H_j \qquad (25)$$

The part in S_k can make no contribution to the matrix elements we are calculating now so, if we set

$$\delta\, g_{jk} = g_{jk}^{(2)} - g_{jk}^{(1)} \qquad (26)$$

we can work simply with

$$\mathscr{H}_1 = \beta \sum_{jk} \delta\, g_{jk}\, S_{2k}\, H_j \qquad (27)$$

---

[2]) I thank Dr. A. J. *Stone* for a remark which simplified this step of the proof.

For calculational purposes it is convenient to rewrite this, using Eq (16) and the shift operators $S_2^\pm$, as

$$\mathcal{H}_1 = \beta H \sum_j \delta g_{jz} h_j S_{2z} + \tfrac{1}{2} \beta H \sum_j (\delta g_{jx} - i \delta g_{jy}) h_j S_2^+$$
$$+ \tfrac{1}{2} \beta H \sum_j (\delta g_{jx} + i \delta g_{jy}) h_j S_2^- \qquad (28)$$

whose matrix elements follow from the formulae (Ref. (20) Appendix 7)

$$\langle S - 1\, M | S_{2z} | SM \rangle = -f(S_1 S_2 S)(S^2 - M^2)^{\frac{1}{2}}$$
$$\langle S - 1\, M \pm 1 | S_2^\pm | SM \rangle = \mp\, f(S_1 S_2 S)(S \mp M - 1)^{\frac{1}{2}}(S \mp M)^{\frac{1}{2}} \qquad (29)$$

where

$$f(S_1 S_2 S) = [(S^2 - (S_1 - S_2)^2)((S_1 + S_2 + 1)^2 - S^2)/(4 S^2 (4 S^2 - 1))]^{\frac{1}{2}} \qquad (30)$$

The contribution to the susceptibility is given by *Van Vleck's* formula (Ref. (20), Eq. (5.64)):

$$\delta \chi = -\frac{2 N}{2 S + 1} \sum W_2 \qquad (31)$$

where

$$H^2 \sum W_2 = - \sum_{\substack{M, M' \\ S' = S \pm 1}} \{E(S') - E(S)\}^{-1} |\langle S'M' | \mathcal{H}_1 | SM \rangle|^2 \qquad (32)$$

Using Eqs. (6), (28), (29) and (32) we find that the $M' = M$ contribution to $\sum W_2$ is

$$C = -\beta^2 \Big(\sum_j \delta g_{jz} h_j\Big)^2 \sum_{M=-S}^{S} \left[\frac{f(S_1 S_2 S)^2 (S^2 - M^2)}{-SJ} \right.$$
$$\left. + \frac{f(S_1 S_2 S + 1)^2 ((S+1)^2 - M^2)}{(S+1) J} \right] \qquad (33)$$

$$= (1/3)\, \beta^2 J^{-1} \Big(\sum_j \delta g_{jz} h_j\Big)^2 [(4 S^2 - 1) f(S_1 S_2 S)^2$$
$$- (4(S+1)^2 - 1)\, f(S_1 S_2 S + 1)^2]$$

Writing

$$\varepsilon = \frac{S_1(S_1 + 1) - S_2(S_2 + 1)}{S(S + 1)} \qquad (34)$$

and using Eq. (30), we deduce that

$$C = (1/12) \, \beta^2 \, J^{-1} \, (2S + 1) \, (1 - \varepsilon^2) \, (\sum_j \delta \, g_{jz} h_j)^2 \tag{35}$$

It is clear from symmetry, and it also follows from a direct evaluation, that there are similar contributions refering to δg_{jx} and δg_{jy}. Let us again average over angle and then we get finally

$$\delta \bar{\chi} = - \frac{N\beta^2}{18 \, J} \, (1 - \varepsilon^2) \sum_{jk} (g_{jk}^{(1)} - g_{jk}^{(2)})^2 \tag{36}$$

Before continuing we should note that although our derivation of Eq. (36) is correct for general S there are some special cases which need examination. These are the cases of maximum and minimum S when the states with $S+1$ or $S-1$, respectively, do not exist and, also ε is undefined for $S=0$. It is easy to see that Eq. (36) still holds for these providing that we define $\varepsilon = 2S_1 + 1$ for the case $S_1 = S_2$, $S = 0$.

Putting together our findings so far, we have the following formulae for the susceptibility of the coupled system. For a set of states of given S:

$$\bar{\chi} \, (S_1 S_2 S) = \frac{N\beta^2 \, S(S+1)}{9 \, kT} \sum_{jk} g_{jk}^2 - \frac{N\beta^2 (1 - \varepsilon^2)}{18 \, J} \sum_{jk} (g_{jk}^{(1)} - g_{jk}^{(2)})^2 \tag{37}$$

where ε has been defined above. Then the total susceptibility is

$$\chi = \frac{\sum\limits_{S=|S_1-S_2|}^{S_1+S_2} (2S + 1) \chi (S_1 S_2 S) \exp[- JS(S + 1)/2kT]}{\sum\limits_{S=|S_1-S_2|}^{S_1+S_2} (2S + 1) \exp[- JS(S + 1)/2kT]} \tag{38}$$

In case both the constituent g-tensors are isotropic, so that $g_{jk}^{(1)} = g_1 \delta_{jk}$ and $g_{jk}^{(2)} = g_2 \delta_{jk}$, then g is given by Eq. (8) and after a little manipulation we see that

$$\bar{\chi} \, (S_1 S_2 S) = \frac{N\beta^2 g^2 S(S + 1)}{3 \, kT} + \frac{2 \, N\beta^2 (g - g_1) \, (g - g_2)}{3 \, J} \tag{39}$$

which is essentially the same as a formula given previously Ref. (20), Eq. (5.73); the derivation was given incorrectly in earlier printings of this reference).

In applying formulae (37) and (38) one would often know the principal g values for the two constituent systems together with the relative orientations of the axes thereof; accordingly it is interesting to restate our results in these terms. Suppose, then, that we have

$$\mathcal{H}_1 = \beta(g_x^{(1)} H_x S_{1x} + g_y^{(1)} H_y S_{1y} + g_z^{(1)} H_z S_{1z})$$

$$\mathcal{H}_2 = \beta(g_{x'}^{(2)} H_{x'} S_{2x'} + g_{y'}^{(2)} H_{y'} S_{2y'} + g_{z'}^{(2)} H_{z'} S_{2z'}) \tag{40}$$

for the two subsystems, where the axes $OX'Y'Z'$ have direction cosines l_{jk} relative to $OXYZ$. Then

$$\mathcal{H}_2 = \beta \sum_{jkm} g_j^{(2)} l_{jk} l_{jm} H_k S_{2m} \tag{41}$$

whence

$$g_{jk}^{(1)} = g_j^{(1)} \delta_{jk},$$

$$g_{jk}^{(2)} = \sum_m g_m^{(2)} l_{mj} l_{mk}, \tag{42}$$

relative to the same set of axes $OXYZ$.

Then for the coupled system

$$g_{jk} = \tfrac{1}{2}(1 - \varepsilon) g_{jk}^{(2)}, j \neq k,$$

$$g_{jj} = \tfrac{1}{2}(1 + \varepsilon) g_j^{(1)} + \tfrac{1}{2}(1 - \varepsilon) g_{jj}^{(2)} \tag{43}$$

whence

$$\sum g_{jk}^{(2)} = \tfrac{1}{4}(1 + \varepsilon)^2 \sum (g_j^{(1)})^2 + \tfrac{1}{2}(1 - \varepsilon^2) \sum g_j^{(1)} g_{jj}^{(2)}$$

$$+ \tfrac{1}{4}(1 - \varepsilon)^2 \sum_{jk} (g_{jk}^{(2)})^2 \tag{44}$$

Now it follows from the fact that the direction cosines form an orthogonal matrix that

$$\sum_{jk} (g_{jk}^{(2)})^2 = \sum_j (g_j^{(2)})^2 \tag{45}$$

Therefore in Eq. (37) we may put

$$\sum_{jk} g_{jk}^2 = \tfrac{1}{4}(1 + \varepsilon)^2 \sum_j (g_j^{(1)})^2 + \tfrac{1}{4}(1 - \varepsilon)^2 \sum_j (g_j^{(2)})^2$$

$$+ \tfrac{1}{2}(1 - \varepsilon^2) \sum_{jk} g_j^{(1)} g_k^{(2)} l_{kj}^2 \tag{46}$$

and

$$\sum_{jk} (g_{jk}^{(1)} - g_{jk}^{(2)})^2 = \sum_j (g_j^{(1)})^2 + \sum_j (g_j^{(2)})^2 - 2\sum_{jk} g_j^{(1)} g_k^{(2)} l_{kj}^2 \qquad (47)$$

When the axes are actually the same, we have $l_{kj} = \delta_{kj}$ and these expressions reduce to

$$\sum_{jk} g_{jk}^2 = \sum_j [\tfrac{1}{2}(g_j^{(1)} + g_j^{(2)}) + \tfrac{1}{2}(g_j^{(1)} - g_j^{(2)})\varepsilon]^2,$$
$$\sum_{jk} (g_{jk}^{(1)} - g_{jk}^{(2)})^2 = \sum_j (g_j^{(1)} - g_j^{(2)})^2. \qquad (48)$$

In the simple but rather important case that the two ions are identical and similarly oriented, so that $g_j^{(1)} = g_j^{(2)}$ we see from Eqs. (37) and (48) that the second term in the susceptibility vanishes identically, thus in this case formula (37) becomes identically zero for the level $S=0$.

A third contribution to the susceptibility rises from matrix elements of the magnetic field operator to states lying outside the coupled set of $(2S_1+1)(2S_2+1)$ states. We now show that this is approximately the same for all states and equal to the sum of the contributions from the two separate systems. Consider system 1. If we neglect energies due to spin-orbit, spin-spin coupling, *etc.*, we can write the excited states as $|a\,S_1'M_1'\rangle$ and because of the form of the magnetic field operator $\mathscr{H}_1 = \beta\mathbf{H}\cdot(\mathbf{L}_1+2\mathbf{S}_1)$ the only non-zero matrix elements to the ground states arise from the part in $\mathbf{L}_1$ and satisfy $S_1'=S_1$, $M_1'=M_1$. In fact

$$\langle S_1 M_1 | \mathscr{H}_1 | a S_1' M_1' \rangle = f(a)\,\delta_{S_1 S_1'}\,\delta_{M_1 M_1'} \qquad (49)$$

where $f(a)$ is independent of the M_1 value (see for example Ref. (23), Chapt. 3). If the set $|a S_1 M_1\rangle$ lies at energy $E(a)$ above the ground state, then the contribution to the susceptibility of the state $|S_1 M_1\rangle$ is given by second-order perturbation theory to be

$$\delta\chi = -2N \sum W_2 = 2N \sum f(a)^2/E(a) \qquad (50)$$

This is the same for each value of M_1.

Now consider a coupled state $|S_1 S_2 SM\rangle$. Through the operator $\mathscr{H}_1$ we get matrix elements to states $|a S_1 S_2 M_1 M_2\rangle$, where system 1 but not system 2 is excited. Because of the selection rules on $\mathscr{H}_1$, when we expand the coupled state we just get the one contribution

$$\langle S_1 S_2 SM | \mathscr{H}_1 | a S_1 S_2 M_1 M_2 \rangle = f(a)\langle S_1 S_2 SM | S_1 S_2 M_1 M_2 \rangle \qquad (51)$$

Hence for the coupled state the contribution of $\mathscr{H}_1$ to χ is given by

$$\delta\chi' = - 2N \Sigma W_2 = 2N \underset{M_1 M_2}{\Sigma} f(a)^2 \langle S_1 S_2 SM | S_1 S_2 M_1 M_2 \rangle^2 / E(a) = \delta\chi \tag{52}$$

This is the same for all S and M. Note, however, that our proof is only approximate as we have neglected the effect of the interaction upon the values of $E(a)$, assuming in Eq. (52) that the $E(a)$ are independent of S and the same as those occuring in Eq. (50). Such effects should usually be small, however, because the excited states of systems 1 and 2 will normally lie at energies large compared with the splittings produced by the interaction.

These last results show that, for example, although the ground $S=0$ state of an antiferromagnetically coupled pair of octahedrally-coordinated Ni^{2+} or Cr^{3+} ions get no contribution though Eq. (37), we do now get a contribution of $\chi = 8N\beta^2/\varDelta$ per ion, where $\varDelta$ is the ligand field splitting parameter (see Ref. (20), pp. 280—281). However there is no such contribution for two high spin ferric or manganous ions.

Finally note that we have neglected the possibility of any fine-structure splitting of the states of either system 1 or system 2. Such splittings will modify the susceptibility at low enough temperatures and we give some discussion of this in section 4.

D. Two-iron Two-sulphur Proteins

In these proteins, such as adrenodoxin or spinach ferredoxin, there are two iron atoms and two acid-labile sulphide ions. The sulphide ions probably occur as bridges between the iron atoms. There are two known oxidation states and it is now generally believed that in the oxidized form there are two antiferromagnetically coupled high-spin ferric ions, giving an $S=0$ state lying lowest, while in the reduced form a high-spin ferric and a high-spin ferrous ion are similarly coupled to give an $S=\frac{1}{2}$ state lowest (3, 4, 22).

Evidently the magnetic susceptibility can give evidence on the magnitude of the coupling between the ions. Except in very accurate work we can simply take each g value to be equal to 2. Write $x = \exp(-J/2\,kT)$ and we get

$$\mu^2 = \frac{24x^2 (1 + 5x^4 + 14x^{10} + 30x^{18} + 55x^{28})}{1 + 3x^2 + 5x^6 + 7x^{12} + 9x^{20} + 11x^{30}} \tag{53}$$

for the oxidized form and

$$\mu^2 = \frac{3 (1 + 10x^3 + 35x^8 + 84x^{15} + 165x^{24})}{1 + 2x^3 + 3x^8 + 4x^{15} + 5x^{24}} \tag{54}$$

for the reduced form. *Palmer, Dunham, Fee, Sands, Iizuka* and *Yonetani* (private communication) have measured the susceptibility of spinach ferredoxin in the range 77°—250 °K and find a best fit to their results by taking $J = 366$ cm^{-1} for the oxidized and $J = 198$ cm^{-1} for the reduced. They assume, as do Eq. (53) and (54), the energy expression given in Eq. (5). As we shall see in section 3B, this assumption may not be entirely accurate. However measurements on adrenodoxin give no evidence for any occupation of any states other than the $S = 0$ and $S = \frac{1}{2}$ ones, up to at least 260 °K, and in this protein J must be at least considerably greater than 300 cm^{-1} (25, 26).

An EPR signal centred at about $g = 1.94$ is observed from the reduced iron-sulphur complex (27) and poseed a considerable problem in its interpretation until the suggestion came that one was dealing with a coupled pair having $S_1 = \frac{5}{2}$, $S_2 = 2$ and a large positive J (22, 24). From Eqs. (8) and (15) it follows that the g values and hyperfine constants for the coupled system are given by

$$g_{jk} = \tfrac{7}{3} g_{jk}^{(1)} - \tfrac{4}{3} g_{jk}^{(2)} ,$$
$$A_{jk}^{(1)} = \tfrac{7}{3} a_{jk}^{(1)} ,$$
$$A_{jk}^{(2)} = - \tfrac{4}{3} a_{jk}^{(2)} ,$$

(55)

in terms of those for the individual ferric and ferrous ions (the relation for the hyperfine constants is incorrect in Ref. (3)).

Gibson et al. (22) measured the three principal g-values for the coupled system to be $g_x = 1.88$, $g_y = 1.94$, $g_z = 2.04$ and assumed $g^{(1)}$ to be isotropic and equal to 2.019, corresponding to a measurement by Title (28) for Fe^{3+} interstitially at an unspecified site in ZnS. With these values, we can deduce from Eq. (55) that $g_x^{(2)} = 2.12$, $g_y^{(2)} = 2.08$ and $g_z^{(2)} = 2.00$. As they point out, these latter g-values fit satisfactorily with the hypothesis that the electron outside the filled half shell in the ferrous ion lies in a d_{z^2} type of orbital.

The hyperfine constants have now been measured by *Sands, Fritz* and *Fee* (see Ref. (3); also for discussion, which is not greatly affected by the numerical error mentioned above). Using Eq. (55) they give, in units 10^{-4} cm^{-1},

$$a_x^{(1)} = - 6.9,\ a_y^{(1)} = - 7.3,\ a_z^{(1)} = - 5.9$$
$$a_x^{(2)} = - 3.4,\ a_y^{(2)} = - 4.8,\ a_z^{(2)} = - 8.5$$

(56)

for the individual ions. These values can be reasonably assigned mainly to the isotropic (Fermi) hyperfine interaction, with a small unexplained anisotropy for the ferric ion and a larger anisotropy for the ferrous ion due mainly to the d_{z^2} electron. For the latter we have, theoretically,

$$a_x^{(2)} = P(g_{Lx}^{(2)} - \varkappa + \tfrac{1}{14})$$
$$a_y^{(2)} = P(g_{Ly}^{(2)} - \varkappa + \tfrac{1}{14}) \tag{57}$$
$$a_z^{(2)} = P(g_{Lz}^{(2)} - \varkappa - \tfrac{1}{7})$$

in the usual terminology (cf Ref. (20), p. 328).

If we now fit $a_x^{(2)} + a_y^{(2)}$ and $a_z^{(2)}$ to these formulae, we get $-P\varkappa = -6.55$ and have as "calculated" values

$$a_x^{(2)} = -3.85, \; a_y^{(2)} = -4.42, \; a_z^{(2)} = -8.5 \tag{58}$$

in the same units as before. The fit is not perfect, one possible explanation[3] being that the d orbital is not strictly d_{z^2}. However if we suppose the odd electron is in a $d_{x^2-y^2}$ type of orbital, the signs of $\tfrac{1}{14}$ and $\tfrac{1}{7}$ are changed in Eq. (57) and it becomes impossible to obtain even an approximate fit. Taken all in all, the measured values of the hyperfine constants give strong confirmation of the general correctness of *Gibson, Hall, Thornley* and *Whatley's* model (22).

3. Three-spin Systems

A. Elementary Discussion

Suppose we have three ions, each without spatial degeneracy. Then in order to give a theoretical formula for the susceptibility we need to know the relative energies of the states of the total system in the absence of the magnetic field, and the susceptibility of each set of $(2S+1)$ states having each allowed total spin S. In many cases the latter will be very close to the spin-only value as, for example, in Cr^{3+}, high spin Mn^{2+}, Fe^{3+} and Cr^{2+} and in some other cases like the nickel acetylacetonate trimer is well given by the extended spin only formula of Eq. (24). Then the total susceptibility is given by

$$\chi = \frac{N\beta^2 g^2}{3kT} \frac{\Sigma S(S+1)(2S+1)e^{-E(S)/kT}}{\Sigma(2S+1)e^{-E(S)/kT}} \tag{59}$$

---

[3] The attempt to assign the entire effect to "orbital reduction" of the last terms in Eq. (57) does not lead to very sensible results.

where the sum is over all total spins S allowing for the fact that there are usually several sets of states for each S. In this article we shall only discuss the problem of determining the energies $E(S)$ which need to be put into the formula (59) and leave the question of how to proceed when the g values of the individual ions differ significantly one from another.

First we ask how many times each spin S occurs in the coupled system. This may be done by first coupling the first spin S_1 to the second S_2, which gives $S' = S_1 + S_2, \ldots, |S_1 - S_2|$. Then couple each S' to S_3 to give $S = S' + S_3, \ldots |S' - S_3|$. For example, for three Ni^{2+} ions we have $S' = 2,1,0$, and then $S = 3,2,1,2,1,0,1$. This may be written succinctly as $S = 3, 2^2, 1^3, 0$. In case $S_1 = S_2 = S_3$ we have the following total spins:

$S_1 = \quad \frac{1}{2}: S = (3/2), \frac{1}{2}^2$

$S_1 = \quad 1: S = 3, 2^2, 1^3, 0$

$S_1 = 3/2: S = (9/2), (7/2)^2, (5/2)^3, (3/2)^4, \frac{1}{2}^2$

$S_1 = \quad 2: S = 6, 5^2, 4^3, 3^4, 2^5, 1^3, 0$

$S_1 = 5/2: S = (15/2), (13/2)^2, (11/2)^3, (9/2)^4, (7/2)^5, (5/2)^6, (3/2)^4, \frac{1}{2}^2$

and any other case is easily worked out.

Now suppose the interaction Hamiltonian is the immediate generalization of Eq. (3), namely

$$\mathscr{H} = J_{12} S_1 \cdot S_2 + J_{23} S_2 \cdot S_3 + J_{31} S_3 \cdot S_1 \qquad (60)$$

Write $S' = S_1 + S_2$ and $S = S_1 + S_2 + S_3$. Then from a theoretical standpoint we have three distinct cases. First suppose $J_{12} = J_{23} = J_{31} = J$. Then

$$\mathscr{H} = J \sum_{i<j} S_i \cdot S_j = \frac{1}{2} J S^2 - \frac{1}{2} J S_1^2 - \frac{1}{2} J S_2^2 - \frac{1}{2} J S_3^2 \qquad (61)$$

from which it follows that

$$E(S) = \frac{1}{2} J [S(S+1) - S_1(S_1+1) - S_2(S_2+1) - S_3(S_3+1)] \qquad (62)$$

Thus the energy depends only on S and the relative energies still satisfy the Landé interval rule given in Eq. (6).

The second case is when two of the coupling constants are the same, say $J_{23} = J_{31}$, but not the third. Here again we are fortunate in that we may write the Hamiltonian in the form (21):

$$\mathscr{H} = J_{12} S_1 \cdot S_2 + J_{23} S' \cdot S_3$$

whence it follows that

$$E(S) = \tfrac{1}{2} J_{12} [S'(S'+1) - S_1(S_1+1) - S_2(S_2+1)]$$
$$+ \tfrac{1}{2} J_{23} [S(S+1) - S'(S'+1) - S_3(S_3+1)] \tag{63}$$

This is also very simple, although now $E(S)$ depends on S' as well as on S.

The third case is when all the coupling constants are unequal. There is no longer any general elementary method for obtaining the energies, although in any particular case the complete matrix of $\mathcal{H}$ can be calculated by actually writing out the coupled states and determining the matrix elements by elementary operations with angular momenta (2). This means that we expand any coupled state in terms of the basic states $|M_1 M_2 M_3\rangle$, where M_1, M_2 and M_3 are the S_{1z}, S_{2z} and S_{3z} values. As a trivial example, if $S_1 = S_2 = S_3 = \tfrac{1}{2}$, the unique $S = M = 3/2$ state is $|\tfrac{1}{2}\tfrac{1}{2}\tfrac{1}{2}\rangle$ and

$$\boldsymbol{S}_1 \cdot \boldsymbol{S}_2 \, |\tfrac{1}{2}\tfrac{1}{2}\tfrac{1}{2}\rangle = (\tfrac{1}{2} S_1^+ S_2^- + \tfrac{1}{2} S_1^- S_2^+ + S_{1z} S_{2z}) \, |\tfrac{1}{2}\tfrac{1}{2}\tfrac{1}{2}\rangle = \tfrac{1}{4} \, |\tfrac{1}{2}\tfrac{1}{2}\tfrac{1}{2}\rangle$$

and similarly for $\boldsymbol{S}_2 \cdot \boldsymbol{S}_3$ and $\boldsymbol{S}_3 \cdot \boldsymbol{S}_1$. Hence

$$E\,(3/2) = \tfrac{1}{4} \, (J_{12} + J_{23} + J_{31}) \tag{64}$$

However, except for very low values of the spins, and also certain special cases, these calculations become very long and tedious and, as we shall see in Sect. 3C, can hardly lead to any very general formulae.

In spite of this, there does exist a general theoretical method for dealing with just this situation of the coupling of three (or more) angular momenta. It is the irreducible tensor method of *Racah* (29) and *Wigner* (30).

For the rest of Sect. 3, we shall depend heavily upon this method and, in particular, will derive general formulae for all the matrix elements of $\mathcal{H}$ for general S_1, S_2 and S_3 (Eq. (82) and (83)).

B. General Energy for Two-spin System

We now return to the two spin system and dicsuss its energies in a manner that forms a suitable basis for the irreducible tensor method. Throughout this section we shall closely follow the usage of *Fano* and *Racah's* book (29), which will henceforth be referred to as FR.

First, however, we draw attention to the fact that the scalar product interaction operator $\boldsymbol{S}_1 \cdot \boldsymbol{S}_2$ is by no means the only possible one which commutes with the total spin S. In particular the somewhat more general operator

$$\mathscr{H}_c = J\,(S_1 \cdot S_2) - j\,(S_1 \cdot S_2)^2 \tag{65}$$

also has this property. In a study on pairs of Mn^{2+} ions, interstitially in MgO, *Harris* and *Owen* (31) measured the positions of all levels except that having $S = 5$ and found the intervals did not approximate at all well to the interval rule shown in Eq. (6). However, they obtained satisfactory agreement with experiment by using the more general interaction given in Eq. (65). This is shown in Table 1. Note that in this case the second order contribution to the calculated intervals is really quite large, being on average some 27% of the total.

Table 1. *Positions of levels of two interacting Mn^{2+} ions interstitially in MgO, after Harris and Owen (31). Energies in $°K$ relative to the ground state. $I(S) = S^{-1}(E(S) - E(S-1))$. Calculated values assume $J = 14.6$, $j/J = 0.05$*

S	$E(S)$	$I(S)$	Exp.	Calc.
0	0	—	—	—
1	$J + 16\tfrac{1}{2}j$	$J + 16\tfrac{1}{2}j$	28.0 ± 3	26.6
2	$3\,J + 43\tfrac{1}{2}j$	$J + 13\tfrac{1}{2}j$	24.5 ± 2	24.5
3	$6\,J + 69\,j$	$J + 8\tfrac{1}{2}j$	20.0 ± 3	20.8
4	$10\,J + 75\,j$	$J + 1\tfrac{1}{2}j$	17.5 ± 3	15.7
5	$15\,J + 37\tfrac{1}{2}j$	$J - 7\tfrac{1}{2}j$	—	9.1

This means, for example, that when discussing the susceptibility of the ferredoxins mentioned in section 2 D, the possibility that the energies deviate from those given in Eq. (5) should really be taken into account. Actually in that case most of the susceptibility comes from the two lowest spin values for each oxidation state. Therefore for the oxidized form $(S_1 = S_2 = 5/2)$ the measured J is rather a measure of $J + (16\tfrac{1}{2})\,j$. For the reduced form Eq. (65) gives energies relative to $S = \tfrac{1}{2}$ of

$$E\,(3/2) = (1\tfrac{1}{2})\,J + 18\,(3/4)\,j$$
$$E\,(5/2) = 4\,J + 40\,j$$
$$E\,(7/2) = (7\tfrac{1}{2})\,J + 48\,(3/4)\,j$$
$$E\,(9/2) = 12\,J + 24\,j$$

and the measured J is largely to be equated with $(2/3)$ E $(3/2) = \mathsf{J} +$ $(12\frac{1}{2})$ j.

It is evident that we could add third or higher order terms to $\mathscr{H}_c$. However from general experience with perturbation theory one might well expect the second order term to be the most important and there is some calculational evidence both for this and for the expectation that j may usually be positive (32).

We now consider this matter from the point of view of the irreducible tensor method. It is convenient to work in terms of the unit contra-standard tensor operators which are defined for each system i by the equation

$$\langle S_i \| u_i^k \| S_i \rangle = 1 \tag{66}$$

Here k is the degree of the tensor operator. In particular, the qth component of the spin for system i is given by

$$S_{iq} = [- S_i (S_i + 1) (2 S_i + 1)]^{\frac{1}{2}} u_{iq}^1 \tag{67}$$

and hence the scalar product is

$$\boldsymbol{S_i \cdot S_j} = - \mathrm{D_1} \boldsymbol{u_i^1 \cdot u_j^1} \tag{68}$$

where

$$\mathrm{D_1} = [S_i(S_i + 1) (2 S_i + 1) S_j(S_j + 1) (2 S_j + 1)]^{\frac{1}{2}} \tag{69}$$

The only combinations of products of components u_{1q}^k with $u_{2q'}^{k'}$ which commute with $\boldsymbol{S}$ are the scalar products and so we contemplate writing a general interaction Hamiltonian in the form

$$\mathscr{H} = \sum_k a_k \boldsymbol{u_1^k \cdot u_2^k} \tag{70}$$

It is easy to see that in this sum there is no point in taking $k > 2$ min. (S_1, S_2) because all matrix elements of such terms vanish because of the properties of the $\bar{V}$ coefficients. Using Eq. (70) and FR Eq. (15.6) we find that the energy of the state with spin S is

$$\mathsf{E}\,(S) = \sum_k a_k (-1)^{S_1 + S_2 + S + k} \bar{\mathrm{W}} \begin{pmatrix} S_1 S_1 k \\ S_2 S_2 S \end{pmatrix} \tag{71}$$

But this equation may be inverted by multiplying through by

$$(-1)^S (2 S + 1) \bar{\mathrm{W}} \begin{pmatrix} S_1 S_1 l \\ S_2 S_2 S \end{pmatrix}$$

and summing over all allowed S. Then using FR Eq. (11.15) we get

$$a_l = (-1)^{S_1+S_2+l}(2\,l+1)\,\sum_S (-1)^S(2\,S+1)\,\overline{W}\begin{pmatrix} S_1 S_1 l \\ S_2 S_2 S \end{pmatrix} \mathsf{E}\,(S) \quad (72)$$

for each $l = 0, 1, 2, \ldots, 2$ min. (S_1, S_2). Conversely substituting these a_l in Eq. (71) gives back the energies $\mathsf{E}(S)$.

Now the most general possible interaction which commutes with the total spin can, in its effect upon our set of $(2\,S_1+1)(2\,S_2+1)$ states, be specified by giving the energy $\mathsf{E}(S)$ for each allowed S. But we have proved that the parameters a_l of Eq. (72) will give these $\mathsf{E}(S)$. Hence $\mathscr{H}$ of Eq. (70), with $k = 0, 1, 2, \ldots, 2$ min. (S_1, S_2) gives an expression for such a most general interaction.

We now complete the relation between our new formalism and the conventional expression $\mathscr{H}_c$ given in Eq. (65). First note that the term with $k = 0$ in $\mathscr{H}$ gives a contribution which is the same for all states. We shall usually set $a_0 = 0$, which means in fact that the centre of gravity of the whole set is zero, because we are only interested in relative energies. Also, for the moment, set $a_k = 0$ for $k > 2$. Then writing

$$\eta = S(S+1) - S_1(S_1+1) - S_2(S_2+1)$$
$$D_2 = D_1[(2\,S_1-1)(2\,S_1+3)(2\,S_2-1)(2\,S_2+3)]^{\frac{1}{2}} \quad (73)$$

where D_1 is given in Eq. (69), we put the explicit formulae which are available (33) for $\overline{W}$ when $k = 1$ or 2 into Eq. (71) to obtain

$$\mathsf{E}\,(S) = -\tfrac{1}{2}D_1^{-1}\,a_1\,\eta + \tfrac{1}{2}D_2^{-1}\,a_2\,[3\,\eta\,(\eta+1) - 4\,S_1(S_1+1)\,S_2(S_2+1)] \quad (74)$$

which must be compared with

$$\langle S_1 S_2 SM | \mathscr{H}_c | S_1 S_2 SM \rangle = \tfrac{1}{2}\,\mathsf{J}\,\eta - \tfrac{1}{4}\,\mathsf{j}\,\eta^2 \quad (75)$$

Apart from the centre of gravity, which is not zero in the latter case, these formulae match if we take

$$a_1 = -D_1(\mathsf{J} + \tfrac{1}{2}\,\mathsf{j}), \quad a_2 = -(1/6)\,\mathsf{j}\,D_2 \quad (76)$$

This means that if we wish to know the matrix of relative energies given by the conventional Hamiltonian $\mathscr{H}_c$ we may use the irreducible tensor expression $\mathscr{H}$, with $k = 1, 2$, providing we substitute for a_1 and a_2 according to Eq. (76). Of course for the simpler Hamiltonian given in Eq. (60) we need only take the term with $k = 1$, thus setting $j = 0$.

We notice that the inclusion of a non-zero j leads also to an alteration of the value of a_1. It is worth mentioning that this is a reflection of an inferior feature of an expansion in terms of $(S_1 \cdot S_2)^k$ as opposed to the expansion given in Eq. (70). If in the former expansion we make a least squares fit to the experimental data (for all S) with a limited number of k, and then add extra terms, the new least squares fit will generally alter all the previously determined coefficients. However when, as in Eq. (71), we have a series of orthogonal functions $\overline{W}$ (see FR Eq. (11.15)) each coefficient a_k is determined independently of all the others providing we give the correct weight $(2\,S + 1)$ to each level in our fitting procedure. The actual value of a_k obtained in such a way is identical with the formula given in Eq. (72). For details of this aspect of the method of least squares see, for example, *Aitken* (34), p. 115—6.

C. Matrix Elements for the Three-spin System

Here we have for the most general Hamiltonian of the kind given in Eq·
(1):

$$\mathscr{H} = \sum_k \sum_{i<j} a_k(i, j)\, u_i^k \cdot u_j^k \tag{77}$$

It is convenient to consider separately the matrix elements M_{ij} of $u_i^k \cdot u_j^k$ for each of the three pairs of values of (i, j).

We shall use the $S_1 S_2(S') S_3 SM$ coupling scheme throughout. Then the simplest matrix is the (1, 2) part. Using FR Eq. (15,6) and (15.7) we get

$$
\begin{aligned}
M_{12} &= \langle S_1 S_2(S')\, S_3 SM \,|\, u_1^k \cdot u_2^k \,|\, S_1 S_2(S'')\, S_3 SM \rangle \\
&= \delta_{S'S''} \langle S_1 S_2 S'M' \,|\, u_1^k \cdot u_2^k \,|\, S_1 S_2 S'M' \rangle \\
&= \delta_{S'S''} (-1)^{S_1+S_2+S'+k}\, \overline{W} \begin{pmatrix} S_1 S_1 k \\ S_2 S_2 S' \end{pmatrix}
\end{aligned}
\tag{78}
$$

Note that for given S_1, S_2, S_3, the matrix is completely specified by giving it as a function of S', S'' for each allowed value of S. I also mention that in actual calculations having only $k = 1$ and 2 it is usually easier to evaluate M_{12} by noting that it is diagonal in S' and the diagonal elements are given by Eq. (75) if we replace S in that equation by S' and correct the centre of gravity.

Next, for the (2, 3) part we use FR Eq. (15.6) and (15.7') to obtain

$$
\begin{aligned}
M_{23} &= \langle S_1 S_2(S')\, S_3 SM \,|\, u_2^k \cdot u_3^k \,|\, S_1 S_2(S'')\, S_3 SM \rangle \\
&= (-1)^{S''+S_3+S+k} \langle S_1 S_2 S' \,\|\, u_2^k \,\|\, S_1 S_2 S'' \rangle\, \overline{W} \begin{pmatrix} S'S''k \\ S_3 S_3 S \end{pmatrix} \\
&= (-1)^{S_1+S_2+S_3+S+S'+S''} (2\,S' + 1)^{\tfrac12} (2\,S'' + 1)^{\tfrac12}
\end{aligned}
\tag{79}
$$

$$
\overline{W} \begin{pmatrix} S'S''k \\ S_2 S_2 S_1 \end{pmatrix} \overline{W} \begin{pmatrix} S'S''k \\ S_3 S_3 S \end{pmatrix}
$$

Finally, in a similar manner:

$$M_{13} = \langle S_1 S_2 (S') S_3 SM \mid u_1^k \cdot u_3^k \mid S_1 S_2 (S'') S_3 SM \rangle$$

$$= (-1)^{S_1 + S_2 + S_3 + S + 2S''} (2S' + 1)^{\frac{1}{2}} (2S'' + 1)^{\frac{1}{2}} \tag{80}$$

$$\overline{W} \begin{pmatrix} S'S''k \\ S_1 S_1 S_2 \end{pmatrix} \overline{W} \begin{pmatrix} S'S''k \\ S_3 S_3 S \end{pmatrix}$$

Note that both M_{23} and M_{13} are zero unless $|S' - S''| \leqslant k$. In the important case that $S_1 = S_2$ we see that

$$M_{23} = (-1)^{S' - S''} M_{13} \tag{81}$$

It follows from this that, if $a_k(2,3) = a_k(1,3)$, the part of $\mathcal{H}$ which has that value for k has all matrix elements zero for which $S' - S''$ is odd.

Eqs. (78—80) give a complete general solution for the determination of the matrix elements of $\mathcal{H}$, as given in Eq. (77). When combined with Eq. (76) they give a similar general solution to the more restricted problems based on the scalar product Hamiltonian of Eq. (60) or the more extended Hamiltonian of Eq. (65) and in any particular case one need merely insert the relevant values of the $\overline{W}$ coefficients from suitable Tables (35). For $k \geqslant 2$ this is usually the simplest way to proceed but for the specially important case of $k = 1$ it is probably easier to substitute for the $\overline{W}$ the algebraic formulae given by *Edmonds* (33) and to use the resulting expressions.

With the Hamiltonian of Eq. (60) we only have non-zero matrix elements for $S' - S'' = 0, \pm 1$. It suffices therefore to give the two formulae ($J_{13} = J_{31}$):

$$\mathcal{H}_{S'S'} = \tfrac{1}{2} J_{12}[S'(S' + 1) - S_1(S_1 + 1) - S_2(S_2 + 1)]$$

$$+ \tfrac{1}{4} \left[(J_{13} + J_{23}) + (J_{13} - J_{23}) \frac{S_1(S_1 + 1) - S_2(S_2 + 1)}{S'(S' + 1)} \right] \tag{82}$$

$$[S(S + 1) - S_3(S_3 + 1) - S'(S' + 1)]$$

and

$$\mathcal{H}_{S', S'-1} = \mathcal{H}_{S'-1, S'} = \frac{J_{23} - J_{13}}{4S'} [(S + S_3 + S' + 1)(S' + S_3 - S)$$

$$(S + S' - S_3)(S + S_3 - S' + 1)(S' + S_1 + S_2 + 1)(S' + S_2 - S_1)$$

$$(S' + S_1 - S_2)(S_1 + S_2 - S' + 1) / (4S'^2 - 1)]^{\frac{1}{2}} \tag{83}$$

The complexity of these formulae, especially of the second, is the justification of the remark made in section 3A that they could hardly be obtained by elementary manipulations and they give a good demonstration of the great power of the irreducible tensor method.

As examples we shall give the complete matrices for the Hamiltonian based on Eq. (65) for $S_1 = S_2 = S_3 = \frac{1}{2}$, 1 and $1\frac{1}{2}$, without assuming that the interaction between different pairs of ions is the same (in agreement with *Sinn* (2) for the energies based on the J parts of the matrices for $S_1 = \frac{1}{2}$ and 1). First for $S_1 = \frac{1}{2}$, 2 min. $(S_1, S_1) = 1$ and so there is only a $k = 1$ contribution, *i.e.* J but no j. We already found $E(S = 3/2)$ in Eq. (64) and the matrix for $S = \frac{1}{2}$ is:

$$\mathscr{H}_{00} = -(3/4)\, J_{12}, \; \mathscr{H}_{11} = -\tfrac{1}{2} J_{23} - \tfrac{1}{2} J_{13} + \tfrac{1}{4} J_{12},$$

$$\mathscr{H}_{10} = \mathscr{H}_{01} = \tfrac{\sqrt{3}}{4}\, (J_{23} - J_{13})$$

When $S_1 = 1$, as in three nickel(II) ions, we find:

$$E\,(S = 0) = -J_{12} - J_{23} - J_{31} + (1/3)\,(j_{12} + j_{23} + j_{13}) ,$$

$$E\,(S = 3) = \quad J_{12} + J_{23} + J_{31} + (1/3)\,(j_{12} + j_{23} + j_{13}) ,$$

and for $S = 2$:

$$\mathscr{H}_{11} = -J_{12} + \tfrac{1}{2} J_{13} + \tfrac{1}{2} J_{23} + (1/3)\,(j_{12} + j_{23} + j_{13}),$$

$$\mathscr{H}_{22} = J_{12} - \tfrac{1}{2} J_{13} - \tfrac{1}{2} J_{23} + (1/3)\,(j_{12} + j_{23} + j_{13}),$$

$$\mathscr{H}_{12} = \tfrac{\sqrt{3}}{2}\, (J_{23} - J_{13}),$$

and for $S = 1$:

$$\mathscr{H}_{00} = -2\, J_{12} - (8/3)\, j_{12},$$

$$\mathscr{H}_{11} = -J_{12} - \tfrac{1}{2} J_{13} - \tfrac{1}{2} J_{23} + \tfrac{1}{3} j_{12} - \tfrac{2}{3}\,(j_{23} + j_{13}),$$

$$\mathscr{H}_{22} = J_{12} - \tfrac{3}{2} J_{13} - \tfrac{3}{2} J_{23} + \tfrac{1}{3} j_{12} - \tfrac{4}{3}\,(j_{23} + j_{13}),$$

$$\mathscr{H}_{10} = \tfrac{2}{\sqrt{3}}\, (J_{23} - J_{13}) + \tfrac{1}{\sqrt{3}}\,(j_{23} - j_{13}),$$

$$\mathscr{H}_{20} = -\tfrac{\sqrt{5}}{3}\,(j_{23} + j_{13}),$$

$$\mathscr{H}_{21} = \tfrac{\sqrt{5}}{2\sqrt{3}}\,(J_{23} - J_{13}) + \tfrac{\sqrt{5}}{\sqrt{3}}\,(j_{23} - j_{13}).$$

Finally for $S_1 = S_2 = S_3 = \frac{3}{2}$, as in three chromium(III) ions we get:

$$\mathsf{E}\,(S = \tfrac{9}{2}) = \tfrac{9}{4}\,(\mathsf{J}_{12} + \mathsf{J}_{23} + \mathsf{J}_{13}) - \tfrac{3}{8}\,(j_{12} + j_{23} + j_{13}).$$

When $S = \frac{7}{2}$:

$$\mathscr{H}_{22} = -\tfrac{3}{4}\,\mathsf{J}_{12} + \tfrac{3}{2}\,\mathsf{J}_{13} + \tfrac{3}{2}\,\mathsf{J}_{23} + \tfrac{33}{8}\,j_{12} + \tfrac{3}{4}\,j_{13} + \tfrac{3}{4}\,j_{23},$$

$$\mathscr{H}_{33} = \tfrac{9}{4}\,\mathsf{J}_{12} - \tfrac{3}{8}\,j_{12} + 3\,j_{13} + 3\,j_{23},$$

$$\mathscr{H}_{23} = \tfrac{3\sqrt{3}}{4}\,(\mathsf{J}_{23} - \mathsf{J}_{13}) + \tfrac{9\sqrt{3}}{8}\,(j_{13} - j_{23}).$$

When $S = \frac{5}{2}$:

$$\mathscr{H}_{11} = -\tfrac{11}{4}\,\mathsf{J}_{12} + \tfrac{3}{4}\,\mathsf{J}_{13} + \tfrac{3}{4}\,\mathsf{J}_{23} - \tfrac{23}{8}\,j_{12} + \tfrac{39}{40}\,j_{13} + \tfrac{39}{40}\,j_{23}.$$

$$\mathscr{H}_{22} = -\tfrac{3}{4}\,\mathsf{J}_{12} - \tfrac{1}{4}\,\mathsf{J}_{13} - \tfrac{1}{4}\,\mathsf{J}_{23} + \tfrac{33}{8}\,j_{12} - \tfrac{1}{8}\,j_{13} - \tfrac{1}{8}\,j_{23}.$$

$$\mathscr{H}_{33} = \tfrac{9}{4}\,\mathsf{J}_{12} - \tfrac{7}{4}\,\mathsf{J}_{13} - \tfrac{7}{4}\,\mathsf{J}_{23} - \tfrac{3}{8}\,j_{12} + \tfrac{1}{40}\,j_{13} + \tfrac{1}{40}\,j_{23},$$

$$\mathscr{H}_{12} = \tfrac{3\sqrt{7}}{2\sqrt{5}}\,(\mathsf{J}_{23} - \mathsf{J}_{13}) + \tfrac{3\sqrt{35}}{20}\,(j_{13} - j_{23})$$

$$\mathscr{H}_{13} = -\tfrac{3\sqrt{14}}{5}\,(j_{13} + j_{23}),$$

$$\mathscr{H}_{23} = \tfrac{2\sqrt{2}}{\sqrt{5}}\,(\mathsf{J}_{23} - \mathsf{J}_{13}) - \tfrac{4\sqrt{10}}{5}\,(j_{13} - j_{23}).$$

When $S = \frac{3}{2}$:

$$\mathscr{H}_{00} = -\tfrac{15}{4}\,\mathsf{J}_{12} - \tfrac{75}{8}\,j_{12},$$

$$\mathscr{H}_{11} = -\tfrac{11}{4}\,\mathsf{J}_{12} - \tfrac{1}{2}\,\mathsf{J}_{13} - \tfrac{1}{2}\,\mathsf{J}_{23} - \tfrac{23}{8}\,j_{12} - \tfrac{53}{20}\,j_{13} - \tfrac{53}{20}\,j_{23},$$

$$\mathscr{H}_{22} = -\tfrac{3}{4}\,\mathsf{J}_{12} - \tfrac{3}{2}\,\mathsf{J}_{13} - \tfrac{3}{2}\,\mathsf{J}_{23} + \tfrac{33}{8}\,j_{12} - \tfrac{3}{4}\,j_{13} - \tfrac{3}{4}\,j_{23},$$

$$\mathscr{H}_{33} = \tfrac{9}{4}\,\mathsf{J}_{12} - 3\,\mathsf{J}_{13} - 3\,\mathsf{J}_{23} - \tfrac{3}{8}\,j_{12} - \tfrac{51}{10}\,j_{13} - \tfrac{51}{10}\,j_{23},$$

$$\mathscr{H}_{01} = \tfrac{5\sqrt{3}}{4}\,(\mathsf{J}_{23} - \mathsf{J}_{13}) - \tfrac{5\sqrt{3}}{8}\,(j_{13} - j_{23}),$$

$$\mathscr{H}_{02} = -\tfrac{3\sqrt{5}}{2}\,(j_{13} + j_{23}),$$

$$\mathscr{H}_{03} = 0$$

$$\mathscr{H}_{12} = \tfrac{2\sqrt{15}}{5}\,(\mathsf{J}_{23} - \mathsf{J}_{13}) - \tfrac{4\sqrt{15}}{5}\,(j_{13} - j_{23}),$$

$$\mathscr{H}_{13} = -\tfrac{3\sqrt{21}}{10}\,(j_{13} + j_{23}),$$

$$\mathscr{H}_{23} = \tfrac{3\sqrt{7}}{4\sqrt{5}}\,(\mathsf{J}_{23} - \mathsf{J}_{13}) - \tfrac{27\sqrt{35}}{40}\,(j_{13} - j_{23}),$$

When $S = \frac{1}{2}$:

$$\mathscr{H}_{11} = -\tfrac{11}{4} J_{12} - \tfrac{5}{4} J_{13} - \tfrac{5}{4} J_{23} - \tfrac{23}{8} j_{12} + \tfrac{19}{8} j_{13} + \tfrac{19}{8} j_{23},$$

$$\mathscr{H}_{22} = -\tfrac{3}{4} J_{12} - \tfrac{9}{4} J_{13} - \tfrac{9}{4} J_{23} + \tfrac{33}{8} j_{12} - \tfrac{9}{8} j_{13} - \tfrac{9}{8} j_{23},$$

$$\mathscr{H}_{12} = \tfrac{\sqrt{3}}{2} (J_{23} - J_{13}) - \tfrac{7\sqrt{3}}{4} (j_{13} - j_{23}).$$

D. Permutational Properties

We have seen with the simple Hamiltonian of Eq. (60) that a high degree of degeneracy exists when all the coupling constants J_{ij} are equal, namely all states with a given S have the same energy. Naturally we do not expect all the coupling constants to be the same unless the complex has a high degree of symmetry and, as a particular feature of this, that the three ions are the same. That is that $S_1 = S_2 = S_3$, which we shall assume for most of this section. However we have also seen that it may be necessary to include extra terms in the Hamiltonian. The following general question then arises. If we have a Hamiltonian built up as a polynomial in the components of the three spin vectors S_1, S_2 and S_3 and which is invariant to all permutations of 1, 2 and 3, what degeneracies or other properties necessarily arise in its eigenfunctions as a consequence of this invariance?

The permutation group on three symbols, P_3, has six elements and three classes, namely the unit element I, the three transpositions (12), (23), (31) and the two cycles of order three (123) and (132). Its character table is shown in Table 2.

Table 2. *Character table of the permutation group on three symbols.* T *is a transposition and C a cycle of order three*

	I	3 T	2 C
A_1	1	1	1
A_2	1	—1	1
E	2	0	—1

We define its operations as follows:

$$(12) \, S_{1x} = S_{2x}, \ etc.$$

$$(12) \, |M_1 M_2 M_3\rangle = |M_2 M_1 M_3\rangle$$

In the operation on the ket, system 1 after the transposition gets the z component of spin angular momentum that system 2 had previously, *etc.* This means, for example, that

$$(123)\, S_{1z} = (13)\, (12)\, S_{1z} = (13)\, S_{2z} = S_{2x}$$

but

$$(123)\, |M_1M_2M_3\rangle = (13)\, (12)\, |M_1M_2M_3\rangle = (13)\, |M_2M_1M_3\rangle = |M_3M_1M_2\rangle$$

This definition preserves under permutation all the connections among the $|M_1M_2M_3\rangle$ which are given by the shift operators (of course when M_1 is merely a number multiplying a ket, then $(12)\, M_1 = M_1$ *etc.*).

Assume that all elements of P_3 leave the Hamiltonian $\mathscr{H}$ invariant. Then we can classify the states of the total system both by their S values and by the irreducible representations of P_3. To determine which these latter representations are, we need to know the characters of the transpositions and of the cycles of order three for each pair of values (S_1, S). Consider first the element (12). Because of the behaviour of the Wigner coefficients under interchange of angular momenta (see *e.g.* Ref. *(20)*, p. 24) we have

$$(12)\, |S_1S_1(S')\,S_1SM\rangle = (-1)^{2\,S_1-S'}|S_1S_1(S')\,S_1SM\rangle \qquad (84)$$

Hence

$$\chi\,(12) = \Sigma\,(-1)^{2\,S_1-S'} \qquad (85)$$

The sum is over all S' which can be obtained both by coupling S_1 with S_1 and by coupling S_1 with S. It is then easy to see that

$$
\begin{aligned}
\chi\,((12)) \;&=\; 1 && \text{if } 3\,S_1 \geqslant S \geqslant S_1 \text{ and } (-1)^{3\,S_1-S} = 1, \\
&=\; 0 && \text{if } 3\,S_1 \geqslant S \geqslant S_1 \text{ and } (-1)^{3\,S_1-S} = -1, \\
&=\; (-1)^{S_1-S} && \text{if } S_1 \geqslant S \geqslant 0 \;\; \text{ and } (-1)^{2\,S_1} = 1, \\
&=\; 0 && \text{if } S_1 \geqslant S \geqslant 0 \;\; \text{ and } (-1)^{2\,S_1} = -1,
\end{aligned}
$$

To obtain the character of (123) it is easier to work directly with the states $|M_1M_2M_3\rangle$ and proceed through the M values from $M = 3S_1$ downwards. The differences between successive χ give the characters we want. It follows that:

$$
\begin{aligned}
\chi\,((123)) \;&=\;\; 1 \text{ when } 3\,S_1 - S = 3\,x \\
&=\; -1 \text{ when } 3\,S_1 - S = 3\,x + 1 \\
&=\;\; 0 \text{ when } 3\,S_1 - S = 3\,x - 1,
\end{aligned}
$$

where x is an integer.

The permutational behaviour of states having $S_1 \leqslant \frac{5}{2}$ is shown in Table 3. An interesting feature, which is true in general, is that when $2S_1$ is odd the two states with $S = \frac{1}{2}$ always belong to the E representation of P_3 and hence cannot be split by any Hamiltonian having invariance under this group. We now consider the behaviour of the system when we have the symmetrical Hamiltonian

$$\mathscr{H} = J \sum_{i<l} (\mathbf{S}_i \cdot \mathbf{S}_l) - j \sum_{i<l} (\mathbf{S}_i \cdot \mathbf{S}_l)^2 \tag{86}$$

Table 3. *Permutational behaviour of states for* $S_1 = S_2 = S_3$. *The representation spanned under* P_3 *is given* (Γ) *together with energies* (E) *relative to the states of lowest spin assuming the Hamiltonian given in Eq.* (86)

$S_1 = \frac{1}{2}$				$S_1 = 1\frac{1}{2}$		
S	Γ	E		S	Γ	E
$\frac{1}{2}$	E	0		$\frac{1}{2}$	E	0
$1\frac{1}{2}$	A_1	$1\frac{1}{2}\,\mathrm{J}$		$1\frac{1}{2}$	A_1	$1\frac{1}{2}\,\mathrm{J} - 8\frac{1}{4}\,j$
					A_2	$1\frac{1}{2}\,\mathrm{J} + 3\frac{3}{4}\,j$
$S_1 = 1$					E	$1\frac{1}{2}\,\mathrm{J} - 14\frac{1}{4}\,j$
S	Γ	E		$2\frac{1}{2}$	A_1	$4\,\mathrm{J} - 7\,j$
0	A_2	0			E	$4\,\mathrm{J} + 2\,j$
1	A_1	$\mathrm{J} - 5\,j$		$3\frac{1}{2}$	E	$7\frac{1}{2}\,\mathrm{J} + 3\frac{3}{4}\,j$
	E	$\mathrm{J} - 2\,j$		$4\frac{1}{2}$	A_1	$12\,\mathrm{J} - 3\,j$
2	E	$3\,\mathrm{J}$				
3	A_1	$6\,\mathrm{J}$				

$S_1 = 2$		
S	Γ	E
0	A_1	0
1	A_2	$\mathrm{J} - 13\,j$
	E	$\mathrm{J} - 4\,j$
2	A_1	$3\,\mathrm{J} - 45\,j$
	$2\,E$	$3\,\mathrm{J} - 19\frac{1}{2}\,j \pm \frac{3}{2}\,j\sqrt{193}$
3	A_1	$6\,\mathrm{J} - 6\,j$
	A_2	$6\,\mathrm{J} + 12\,j$
	E	$6\,\mathrm{J} - 24\,j$
4	A_1	$10\,\mathrm{J} - 10\,j$
	E	$10\,\mathrm{J} + 8\,j$
5	E	$15\,\mathrm{J} + 3\,j$
6	A_1	$21\,\mathrm{J} - 21\,j$

Table 3 (continued)

$S_1 = 2\tfrac{1}{2}$

S	Γ	E
$\tfrac{1}{2}$	E	0
$1\tfrac{1}{2}$	A_1	$1\tfrac{1}{2} J + 6\tfrac{3}{4} j$
	A_2	$1\tfrac{1}{2} J - 11\tfrac{1}{4} j$
	E	$1\tfrac{1}{2} J - 29\tfrac{1}{4} j$
$2\tfrac{1}{2}$	A_1	$4 J - 92 j$
	A_2	$4 J - 20 j$
	$2 E$	$4 J - 44 j \pm 12 j \sqrt{19}$
$3\tfrac{1}{2}$	A_1	$7\tfrac{1}{2} J - 62\tfrac{1}{4} j$
	$2 E$	$7\tfrac{1}{2} J - 14\tfrac{1}{4} j \pm 6 j \sqrt{34}$
$4\tfrac{1}{2}$	A_1	$12 J + 12 j$
	A_2	$12 J + 36 j$
	E	$12 J - 24 j$
$5\tfrac{1}{2}$	A_1	$17\tfrac{1}{2} J - 7\tfrac{1}{4} j$
	E	$17\tfrac{1}{2} J + 22\tfrac{3}{4} j$
$6\tfrac{1}{2}$	E	$24 J$
$7\tfrac{1}{2}$	A_1	$31\tfrac{1}{2} J - 56\tfrac{1}{4} j$

For $S = \tfrac{1}{2}$, 1 and $1\tfrac{1}{2}$ the matrices given at the end of section 3C can be used to give the energies. When $S = 2$ or $2\tfrac{1}{2}$ new matrices are required which can be derived from Eq. (78—80) (note comment after Eq. (78)). The resulting energies in each case are given in Table 3 relative to that of the states with lowest total spin S. They are plotted in Fig. 1 as functions of the ratio j/J. We see the rather remarkable feature that, if $S_1 = \tfrac{5}{2}$ and j/J = 0.05 and J > 0, the effect of the second order terms is so large that it forces a change of ground state, three separate $S = \tfrac{5}{2}$ levels going below the pair of $S = \tfrac{1}{2}$ levels. Evidently near this cross-over we may expect complicated and peculiar magnetic behaviour.

We conclude this subsection by asking what happens when the permutational symmetry is lower but still non-trivial. One case is when $S_1 = S_2$, but S_3 is not necessarily the same, and $\mathscr{H}$ is invariant merely to the transposition (12). The corresponding group P_2 has two classes A_1 and A_2 say. The character of (12) was given before and we also notice the "branching rules" for representations of P_3, which are

$$A_1 \to A_1$$
$$A_2 \to A_2$$
$$E \to A_1 + A_2$$

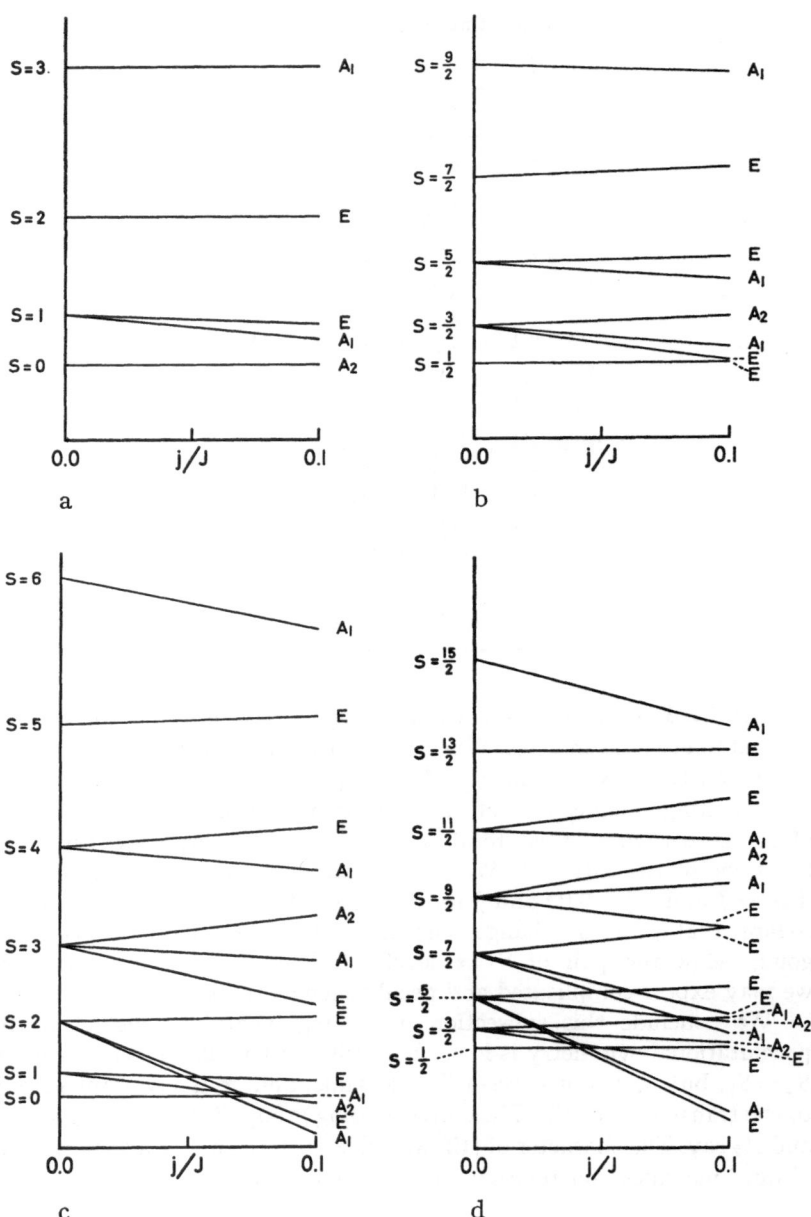

Fig. 1. Energies of systems containing three identical ions, having spin S_1 and interaction Hamiltonian given in Eq. (86), as functions of j/J. (a) $S_1 = 1$, (b) $S_1 = 1\frac{1}{2}$, (c) $S_1 = 2$, (d) $S_1 = 2\frac{1}{2}$

In particular, for the nickel trisacetylacetonate trimer (13) we have under the group P_2:

$$\Gamma(S = 0) = A_2$$
$$\Gamma(S = 1) = 2 A_1 + A_2$$
$$\Gamma(S = 2) = A_1 + A_2$$
$$\Gamma(S = 3) = A_1$$

In the $11(S')1 SM$ coupling scheme for this ion it follows from Eq. (84) that the states with even S' transform according to A_1 while those with odd S' transform according to A_2.

The other case is when $S_1 = S_2 = S_3$ but $\mathscr{H}$ is only invariant to even permutations of the spin vectors. The corresponding permutation group has the three elements 1, (123) and (132). It has the representations shown in Table 4. The fact that two of the representations have complex characters raises special difficulties (see Ref. (20), p. 208) which are best evaded here by still working within the group P_3 (see the end of Sect. 3E). The branching rules here are

$$A_1, A_2 \rightarrow \Gamma_1$$
$$E \rightarrow \Gamma_2 + \Gamma_3 .$$

Table 4. *Character table of the group of even permutations on three symbols.*
$\omega = -\frac{1}{2} + \frac{1}{2}\sqrt{-3}$

	1	(123)	(132)
Γ_1	1	1	1
Γ_2	1	ω	ω^2
Γ_3	1	ω^2	ω

So the main effect of lowering the symmetry is to introduce the possibility of interaction between A_1 and A_2 states.

E. Three Body Interactions

The compound $[Cr_3(CH_3COO)_6(OH)_2]Cl, 8 H_2O$ has a positive $J \approx 20$ cm^{-1} and the three chromic ions lying at the corners of an equilateral triangle. The structure suggests that the Hamiltonian for the spin system should be invariant under P_3 and hence that the two lowest-lying $S = \frac{1}{2}$

J. S. Griffith

states should have the same energy. However *Uryû* and *Friedberg* (36) found a specific heat anomaly at very low temperatures, which they interpreted as arising from a separation between these two doublets of approximately 3 cm⁻¹, and claimed that the apparently symmetrical operator

$$\mathscr{H}_u = J'[(\boldsymbol{S}_1 \cdot \boldsymbol{S}_2)(\boldsymbol{S}_2 \cdot \boldsymbol{S}_3) + (\boldsymbol{S}_2 \cdot \boldsymbol{S}_3)(\boldsymbol{S}_3 \cdot \boldsymbol{S}_1) + (\boldsymbol{S}_3 \cdot \boldsymbol{S}_1)(\boldsymbol{S}_1 \cdot \boldsymbol{S}_2)] \qquad (87)$$

would lead to a separation between the doublets of $3J'\sqrt{3}$. We have already seen that no operator which is invariant under P_3 can do this and therefore there seems to be a contradiction. We now analyse the matter and arrive at a resolution of this.

Let us consider first the general question of how to build polynomials in the three spin vectors, which commute with the total spin $\boldsymbol{S}$. From the discussion given by *Weyl* (37), Sect. II.9) we expect these to be made up of four basic quantities which we shall write

$$
\begin{aligned}
\alpha &= \boldsymbol{S}_2 \cdot \boldsymbol{S}_3 \\
\beta &= \boldsymbol{S}_3 \cdot \boldsymbol{S}_1 \\
\gamma &= \boldsymbol{S}_1 \cdot \boldsymbol{S}_2 \\
\varDelta &= \boldsymbol{S}_1 \cdot (\boldsymbol{S}_2 \wedge \boldsymbol{S}_3)
\end{aligned} \qquad (88)
$$

and of course numbers. We do not need to include such an operator as $\boldsymbol{S}_1^2$ because it is equivalent to the number $S_1(S_1 + 1)$ within the set of states that we are considering. Although each of α, β, γ, $\varDelta$, and any product of them, commutes with $\boldsymbol{S}$, they do not commute one with another but have the following commutators:

$$
\begin{aligned}
[\alpha, \beta] &= [\beta, \gamma] = [\gamma, \alpha] = -i\varDelta, \\
[\alpha, \varDelta] &= i(S_2^2\beta - \gamma S_3^2 - \gamma\alpha + \alpha\beta), \\
[\beta, \varDelta] &= i(S_3^2\gamma - \alpha S_1^2 - \alpha\beta + \beta\gamma), \\
[\gamma, \varDelta] &= i(S_1^2\alpha - \beta S_2^2 - \beta\gamma + \gamma\alpha).
\end{aligned} \qquad (89)
$$

However $\alpha + \beta + \gamma$ does commute with $\varDelta$ as one sees immediately from Eq. (89).

As well as commuting with $\boldsymbol{S}$, a suitable operator must be Hermitian, otherwise the energies of the system will generally not be real. The operator $\mathscr{H}_u$ clearly does not have this property because

$$\mathscr{H}_u - \overline{\mathscr{H}}_u = J'[\gamma, \alpha] + J'[\alpha, \beta] + J'[\beta, \gamma] = -3iJ'\varDelta \qquad (90)$$

116

nor, because $\alpha\beta \neq \beta\alpha$ etc., is it even invariant to all permutations of the group P_3 but only to the even ones. Especially for the first of these reasons, $\mathscr{H}_u$ is not a suitable operator to use. However, from any non-Hermitian $\mathscr{H}_u$ we easily construct the Hermitian operatior

$$\mathscr{H} = \mathscr{H}_u + \overline{\mathscr{H}}_u \tag{91}$$

which is suitable on both counts. Of course, $\mathscr{H}$ cannot split the two ground doublets.

We can now relate the operator $\mathscr{H}_u$ to the second-order operator we discussed previously, namely,

$$K = \sum_{i<j} (S_i \cdot S_j)^2 = \alpha^2 + \beta^2 + \gamma^2 \tag{92}$$

Within our set of states we have:

$$\alpha + \beta + \gamma = \sum_{i<j} (S_i \cdot S_j) = \tfrac{1}{2} S(S+1) - \tfrac{3}{2} S_1(S_1 + 1) \tag{93}$$

Hence

$$\tfrac{1}{4} [S(S+1) - 3 S_1(S_1 + 1)]^2 = K + \mathscr{H}$$

$$\mathscr{H}_u = \tfrac{1}{2} (\mathscr{H}_u + \overline{\mathscr{H}}_u) + \tfrac{1}{2} (\mathscr{H}_u - \overline{\mathscr{H}}_u) = \tfrac{1}{2} \mathscr{H} - \tfrac{3}{2} i\Delta \tag{94}$$

This means that if we know the matrix of K, which we do, and of Δ, we can immediately deduce that of $\mathscr{H}_u$. We discuss how to treat Δ in a moment and anticipate to say that the matrix elements of $\mathscr{H}_u$ within the two ground doublets are given by

S'	1	2
1	$\frac{123}{16} J'$	$\frac{3\sqrt{3}}{2} J'$
2	$-\frac{3\sqrt{3}}{2} J'$	$\frac{123}{16} J'$

This illustrates dramatically the innappropriateness of the operator $\mathscr{H}_u$ because it leads to complex energies! As to what does give rise to the apparent splitting, I do not know, but of course it follows from the matrix elements given in section 3C that a small departure from the assumed equality of the J values would be one possibility.

When it comes to fitting the experimentally measured susceptibility for this compound I would suggest using Eq. (59) with the energies $E(S)$ which are given in Table 3 for $S_1 = 1\frac{1}{2}$, remembering that each E representation must be counted twice in both the sums given in that equation. Because the $S = \frac{1}{2}$ doublets are not split with this model, the very low temperature specific heat behaviour may not fit at all well. Probably the best thing to do about that is to introduce an extra purely phenomenological parameter, δ say, to represent a small splitting of the doublet. For the susceptibility, we should then still use Eq. (59) and the energies of Table 3, except for the ground doublet for which we would take two energies as $E_1(\frac{1}{2}) = 0$ and $E_2(\frac{1}{2}) = \delta$. However as we have thus introduced a small separation between two doublets, each of which has essentially the same susceptibility, it is unlikely that this extra refinement will prove necessary as far as fitting the susceptibility curve is concerned.

The matrix elements of the operator Δ are easily obtained by irreducible tensor methods. Write

$$\mu_i = [S_i(S_i + 1)(2S_i + 1)]^{\frac{1}{2}} \tag{95}$$

Then we easily see that

$$\Delta = i\mu_1\mu_2\mu_3 \sqrt{2} \, (u_1^1 x u_2^1)^1 \cdot u_3^1 \tag{96}$$

in terms of the unit tensor operators defined in Eq. (66). Δ is Hermitian while the matrix elements of $(u_1^1 x u_2^1)^1 \cdot u_3^1$ are all real in our coupling scheme. It follows that the matrix of Δ is skew-symmetric. Actually, using FR Eq. (15.4) and (15.6), we find that a general matrix element is

$$Q_{S'S''} = \langle S_1 S_2(S') S_3 SM \,|\, \Delta \,|\, S_1 S_2(S'') S_3 SM \rangle$$

$$= (-1)^{S''+S_3+S-\frac{1}{2}} \mu_1\mu_2\mu_3 \sqrt{6} \, (2 S' + 1)^{\frac{1}{2}}(2 S'' + 1)^{\frac{1}{2}} \tag{97}$$

$$\text{X} \begin{bmatrix} S_1 S_1 1 \\ S_2 S_2 1 \\ S' S'' 1 \end{bmatrix} \bar{W} \begin{pmatrix} S' S'' 1 \\ S_3 S_3 S \end{pmatrix}$$

$Q_{S'S''}$ is zero if $S' = S''$ or if $|S' - S''| > 1$. Hence it suffices to give its value under the assumption that $S'' = S' - 1$. We can then evaluate X through FR Eq. (12.12) and obtain the following result:

$$Q_{S'S^*} = \tfrac{1}{4}i\,[(S + S_3 + S' + 1)(S' + S_3 - S)(S + S' - S_3)$$
$$(S + S_3 - S' + 1)(S' + S_1 + S_2 + 1)(S' + S_2 - S_1) \qquad (98)$$
$$(S' + S_1 - S_2)(S_1 + S_2 - S' + 1)/(4\,S'^2 - 1)]^{\tfrac{1}{2}}$$

In the particular case that $S_1 = S_2 = S_3$, $S' = S_1 + \tfrac{1}{2}$, $S = \tfrac{1}{2}$ we get

$$Q_{S'S^*} = \tfrac{1}{4}i\,\sqrt{3}\,(S_1 + \tfrac{1}{2})^2 \qquad (99)$$

from which we get a splitting δE of the ground doublets as follows:

$$S_1 = \tfrac{1}{2}, \quad \delta E = \tfrac{1}{2}\sqrt{3}$$
$$S_1 = \tfrac{3}{2}, \quad \delta E = 2\sqrt{3}$$
$$S_1 = \tfrac{5}{2}, \quad \delta E = \tfrac{9}{2}\sqrt{3}$$

However Δ is not a very suitable operator to use directly to split the ground doublets. It is Hermitian and invariant under even permutations of S_1, S_2 and S_3. But its behaviour under the *Kramers*, or time-reversal operator (*) is rather unsatisfactory in that $\Delta^* = -\Delta$.

It is interesting, therefore, to ask if there exists an operator D which satisfies $\bar{D} = D^* = D$ and is invariant only under even permutations of the group P_3. The answer to this is yes, and an example is given by

$$D = \alpha\beta^2 + \beta^2\alpha + \beta\gamma^2 + \gamma^2\beta + \gamma\alpha^2 + \alpha^2\gamma \qquad (100)$$

Such an operator is suitable for a situation in which the three ions are equivalent through the existence merely of a C_3 axis, but is probably not so for our chromic compound because of its higher symmetry. The question arises now, however, whether an operator with the general properties given above for D could split the pair of doublets having $S = \tfrac{1}{2}$.

To discuss this, it is most convenient to work within the group P_3. Then define

$$D^{A_1} = \tfrac{1}{2}\{1 + (12)\}D,$$
$$D^{A_2} = \tfrac{1}{2}\{1 - (12)\}D. \qquad (101)$$

It is immediate that D^{A_1} and D^{A_2} transform according to the irreducible representations A_1 and A_2, respectively, of the group P_3 (which is isomorphic with the dihedral group D_3) and are each separately

Hermitian and unchanged by the * operation. The notation used here conforms with the finite group irreducible tensor scheme $(38, 39)$. We have already proved that D^{A_1} cannot split the doublets and need therefore only discuss D^{A_2}. Now the states concerned are

$$
\begin{aligned}
|a,m\rangle &= |S_1 S_1 (S_1 - \tfrac{1}{2}) S_1 \tfrac{1}{2} m\rangle \\
|b,m\rangle &= |S_1 S_1 (S_1 + \tfrac{1}{2}) S_1 \tfrac{1}{2} m\rangle
\end{aligned}
\tag{102}
$$

and it follows from Eq. (84) that

$$
\begin{aligned}
(12) \, |a,m\rangle &= (-1)^{S_1+\frac{1}{2}} |a,m\rangle \\
(12) \, |b,m\rangle &= (-1)^{S_1-\frac{1}{2}} |b,m\rangle
\end{aligned}
\tag{103}
$$

Hence (for notation see (39) Sect. (3.4)) they form the (c,s) components of the E representation (in some order and not necessarily correctly connected in phase). But for the (c,s) components we have (Ref. (39), p. 119):

$$
\begin{aligned}
\langle Ec \,|\, D^{A_2} \,|\, Ec \rangle &= \langle E \| D^{A_2} \| E \rangle \, V \begin{pmatrix} E\,E\,A_2 \\ c\,c\,\iota \end{pmatrix} \\
&= 0 = \langle Es \,|\, D^{A_2} \,|\, Es \rangle , \\
\langle Ec \,|\, D^{A_2} \,|\, Es \rangle &= \langle E \| D^{A_2} \| E \rangle \, V \begin{pmatrix} E\,E\,A_2 \\ c\,s\,\iota \end{pmatrix} \\
&= -\langle E \| D^{A_2} \| E \rangle \, V \begin{pmatrix} E\,E\,A_2 \\ s\,c\,\iota \end{pmatrix} \\
&= -\langle Es \,|\, D^{A_2} \,|\, Ec \rangle
\end{aligned}
\tag{104}
$$

Hence the matrix of D^{A_2} is skew-symmetric.

The effect of the * operation on the kets of Eq. (102) is obtained by expanding them in terms of $|M_1 M_2 M_3\rangle$ and using Ref. (20), p. 208—9, examples 6 and 7. We find

$$
\begin{aligned}
|a,m\rangle^* &= (-1)^{3S_1-S+m} |a, -m\rangle , \\
\langle a,m|^* &= (-1)^{-3\,S_1+S-m} \langle a, -m|
\end{aligned}
\tag{105}
$$

with similar equation for $|b, m\rangle$. Hence

$$\begin{aligned}
\langle a, m | D^{A_2} | b, m \rangle &= \overline{\langle b, m | D^{A_2} | a, m \rangle} \\
&= \langle b, m | D^{A_2} | a, m \rangle^* \\
&= \langle b, m | D^{A_2} | a, m \rangle
\end{aligned} \tag{106}$$

These are the only matrix elements of D^{A_2} which can be non-vanishing. So we have also proved that the matrix of D^{A_2} is symmetric. Hence it is zero and therefore the operator D cannot split the ground doublets.

4. Effect of Fine-structure Splitting

When some of the J_{ij} are negative, the level of highest spin, namely $S = S_1 + S_2 + S_3$, can lie lowest in energy and hence at very low temperatures the Bohr magneton number μ calculated from Eq. (59) will satisfy

$$\mu^2 = g^2 S (S + 1) \tag{107}$$

per trimeric unit. However, there is now the possibility of a fine structure splitting of this level with a resultant alteration in the magnetic susceptibility at temperatures comparable with the magnitude of this splitting. In fact *Ginsberg, Martin* and *Sherwood* (*13*) investigated the linear trimeric cluster of bis(acetylacetonato)nickel(II) and found $J = -52$ cm^{-1} between adjacent nickel ions. At 4.3 °K the susceptibility is quite close to that given by Eq. (107) with $S = 3$, but below that temperature μ suffers a sharp decrease. *Ginsberg et al.* (*13*) point out that this might be due either to a fine structure splitting of the level or to antiferromagnetic interaction between adjacent trimeric units.

In predominantly ferromagnetically-coupled systems, then, it is of interest to know the effect of a fine-structure splitting upon the susceptibility of the state of highest spin. And because, for large clusters, this spin may be large, it would be useful to possess formulae for general S, if possible. This can probably not be done in any very compact manner for a general fine-structure splitting but in one important special case we can do so, as we shall now see. This case is when the interaction with the magnetic field has the isotropic form $g \beta \mathbf{H} \cdot \mathbf{S}$ and the fine-structure splitting can be represented as DS_z^2 (we omit the centre of gravity correction $-\frac{1}{3} DS(S + 1)$). We derive formulae with the magnetic field along OZ, which gives $\chi_{\parallel}$, and along OX, which gives $\chi_{\perp}$. In a polycrystalline specimen $\chi = \frac{1}{3} \chi_{\parallel} + \frac{2}{3} \chi_{\perp}$.

We use Van Vleck's formula for the susceptibility, namely

$$\chi = N \frac{\Sigma \, (W_1^2/kT - 2 \, W_2) \, e^{-W_0/kT}}{\Sigma \, e^{-W_0/kT}} \tag{108}$$

where the W_0 are DM^2. First let the field H lie along OZ. Then

$$W_1 = g \, \beta M, \quad W_2 = 0$$

where M runs from $-S$ to $+S$. Thus

$$\chi_\| = N \frac{\Sigma \, (g^2 \beta^2 M^2/kT) \, e^{-DM^2/kT}}{\Sigma \, e^{-DM^2/kT}} \tag{109}$$

Set $x = D/kT$ and we then have

$$\mu_\|^2 = 3 \, kT \, \chi_\| / (N\beta^2) = \frac{6 \, g^2 \sum\limits_{M=1}^{S} M^2 e^{-M^2 x}}{\sum\limits_{M=-S}^{S} e^{-M^2 x}}, \; S \text{ integral}$$

$$= \frac{3 \, g^2 \sum\limits_{M=\frac{1}{2}}^{S} M^2 e^{-M^2 x}}{\sum\limits_{M=\frac{1}{2}}^{S} e^{-M^2 x}}, \; S \text{ half-integral}$$

Now let the field lie along OX. The interaction with the magnetic field is now

$$g \, \beta \, HS_x = \tfrac{1}{2} g \, \beta \, H(S^+ + S^-)$$

and hence all the diagonal elements of the magnetic field are zero. At first sight, then, it would seem that all $W_1 = 0$. However there is one exception, namely when S is half-integral and $M = \pm\frac{1}{2}$. Both these states have the same W_0 but have the matrix element between them of ($\pm\frac{1}{2}$ refer to the M value):

$$\langle -\tfrac{1}{2} | \tfrac{1}{2} g \, \beta \, H(S^+ + S^-) | \tfrac{1}{2} \rangle = \tfrac{1}{2} g \, \beta \, H \langle -\tfrac{1}{2} | S^- | \tfrac{1}{2} \rangle = \tfrac{1}{2} g \, \beta \, H(S + \tfrac{1}{2}) \tag{110}$$

Hence for this pair, $W_1 = \pm\frac{1}{2} g \, \beta (S + \tfrac{1}{2})$.

All non-vanishing matrix elements of $g \beta H S_x$ satisfy $\Delta M = \pm 1$ and are given by the general formulae (see Ref. (23), p. 48):

$$\langle M | \tfrac{1}{2} g \beta H (S^+ + S^-) | M + 1 \rangle = \tfrac{1}{2} g \beta H [(S + M + 1)(S - M)]^{\frac{1}{2}}$$
$$\langle M | \tfrac{1}{2} g \beta H (S^+ + S^-) | M - 1 \rangle = \tfrac{1}{2} g \beta H [(S + M)(S - M + 1)]^{\frac{1}{2}}$$

$$(111)$$

These contribute to W_2 for a state with given M through the second order perturbation theory expression [4]

$$W_2 = - \sum_{M'} \frac{|\langle M | g \beta S_x | M' \rangle|^2}{E(M') - E(M)} \tag{112}$$

which is readily seen to reduce to

$$
\begin{aligned}
W_2 &= \frac{g^2 \beta^2 [S(S+1) + M^2]}{2 D (4 M^2 - 1)}, \quad M \neq \pm \tfrac{1}{2} \\
&= - \frac{g^2 \beta^2 (S + 1\tfrac{1}{2})(S - \tfrac{1}{2})}{8 D}, \quad M = \pm \tfrac{1}{2}
\end{aligned}
$$

We can now write down general formulae for $\chi_\perp$ and give the results as follows:

$$\mu_\perp^2 = - 3 g^2 x^{-1} \frac{\Sigma (S(S+1) + M^2)(4 M^2 - 1)^{-1} e^{-M^2 x}}{\Sigma e^{-M^2 x}} \tag{113}$$

when S is an integer. The sum is over all values of M.

When S is half-integral we find:

$$\mu_\perp^2 =$$

$$\frac{\tfrac{3}{4} g^2 (S + \tfrac{1}{2})^2 + 3 g^2 x^{-1} [\tfrac{1}{4}(S + 1\tfrac{1}{2})(S - \tfrac{1}{2}) - \sum\limits_{M=1\frac{1}{2}}^{S} (S(S+1) + M^2)(4 M^2 - 1)^{-1} e^{-(M^2 - \frac{1}{4})}]}{\sum\limits_{M=\frac{1}{2}}^{S} e^{-(M^2 - \frac{1}{4})x}} \tag{114}$$

When $T \to \infty$, i. e. $x \to 0$, all these formulae tend to the normal extended spin-only value $\mu^2 = g^2 S(S+1)$. As $T \to 0$ we get various limiting forms which are most conveniently expressed as

---

[4] Strictly speaking, this is a formula from non-degenerate perturbation theory and should not be used when $M = \pm \tfrac{1}{2}$. However one can show that ΣW_2 is still given correctly by using it, because on replacing $|\pm \tfrac{1}{2}\rangle$ with any orthonormal transform, this sum is unaltered.

$$\chi_\| \to 0$$
$$\chi_\perp \to N \; \beta^2 g^2 S(S+1) \; D^{-1}$$
$$\chi \to \tfrac{2}{3} N \; \beta^2 g^2 S(S+1) \; D^{-1}$$

when S is an integer, and

$$\mu_\|^2 \to \tfrac{3}{4} g^2$$
$$\mu_\perp^2 \to \tfrac{3}{4} g^2 (S + \tfrac{1}{2})^2$$
$$\mu^2 \to \tfrac{1}{8} g^2 (4 S^2 + 4 S + 3)$$

when S is half-integral.

Finally, for easy reference, we give the special forms which the general formulae take when $g = 2$ and $S \leqslant 3$. To obtain the values for $g \neq 2$, multiply by $\tfrac{1}{4} g^2$.

$S = 0$: $\mu_\|^2 = \mu_\perp^2 = \mu^2 = 0$.

$S = \tfrac{1}{2}$: $\mu_\|^2 = \mu_\perp^2 = \mu^2 = 3$.

$S = 1$: $\mu_\|^2 = 24 \, e^{-x} A$, $\mu_\perp^2 = 24 \, x^{-1}(1 - e^{-x}) \, A$,
  $\mu^2 = 8 \, \{2 \, x^{-1} + (1 - 2 \, x^{-1}) \, e^{-x}\} \, A$
  where $A = 1/(1 + 2 \, e^{-x})$.

$S = \tfrac{3}{2}$: $\mu_\|^2 = 3 \, (1 + 9 \, e^{-2x}) \, A$, $\mu_\perp^2 = 3 \, x^{-1}(4 \, x + 3 - 3 \, e^{-2x}) \, A$,
  $\mu^2 = 3 \, (3 + 2 \, x^{-1} + (3 - 2 \, x^{-1}) \, e^{-2x}) \, A$,
  where $A = 1/(1 + e^{-2x})$.

$S = 2$: $\mu_\|^2 = 24 \, (e^{-x} + 4 \, e^{-4x}) \, A$, $\mu_\perp^2 = 8 \, x^{-1}(9 - 7 \, e^{-x} - 2 \, e^{-4x}) \, A$,
  $\mu^2 = 8 \, (6 \, x^{-1} + (1 - \tfrac{14}{3} \, x^{-1}) e^{-x} + (4 - \tfrac{4}{3} \, x^{-1}) e^{-4x}) \, A$,
  where $A = 1/(1 + 2 \, e^{-x} + 2 \, e^{-4x})$.

$S = \tfrac{5}{2}$: $\mu_\|^2 = 3 \, (1 + 9 \, e^{-2x} + 25 \, e^{-6x}) A$, $\mu_\perp^2 = 3 \, (9 + \tfrac{1}{2} x^{-1}$
  $(16 - 11 \, e^{-2x} - 5 \, e^{-6x})) \, A$, $\mu^2 = (19 + 16 \, x^{-1}$
  $+ (9 - 11 \, x^{-1}) e^{-2x} + (25 - 5 \, x^{-1}) e^{-6x}) \, A$,
  where $A = 1/(1 + e^{-2x} + e^{-6x})$.

$S = 3$: $\mu_\|^2 = 24 \, (e^{-x} + 4 \, e^{-4x} + 9 \, e^{-9x}) \, A$,
  $\mu_\perp^2 = 8 \, x^{-1}(18 - 13 \, e^{-x} - \tfrac{16}{5} \, e^{-4x} - \tfrac{9}{5} \, e^{-9x}) \, A$,
  $\mu^2 = 8 \, (12 \, x^{-1} + (1 - \tfrac{26}{3} \, x^{-1}) e^{-x} + (4 - \tfrac{32}{15} \, x^{-1}) \, e^{-4x}$
  $+ (9 - \tfrac{6}{5} \, x^{-1}) \, e^{-9x}) \, A$,
  where $A = 1/(1 + 2 \, e^{-x} + 2 \, e^{-4x} + 2 \, e^{-9x})$.

These special cases have been discussed previously elsewhere. For $S \leqslant 2$, see Ref. (40), which has errors. For $S = \frac{5}{2}$, see (41, 42, 43), the last of which has an error. For $S = 3$, see (7).

Acknowledgement: This work was done partly while the author was a Professeur invité at the University of Geneva.

5. References

1. *Kokoszka, G. F., Duerst, R. W.:* Coord. Chem. Rev. *5*, 209 (1970).
2. *Sinn, E.:* Coord. Chem. Rev. *5*, 313 (1970).
3. *Bearden, A. J., Dunham, W. R.:* Struct. Bonding *8*, 1 (1970).
4. *Tsibris, J. C. M., Woody, R. W.:* Coord. Chem. Rev. *5*, 417 (1970).
5. *Lemberg, M. R.:* Physiol. Rev. *49*, 48 (1969).
6. *Van Gelder, B. F., Beinert, H.:* Biochim. Biophys. Acta *189*, 1 (1969).
7. *Griffith, J. S.:* Mol. Phys. *21*, 141 (1971).
8. *Scheinberg, I.:* In: The Biochemistry of Copper, pp. 513—524; edited by *Peisach, J., Aisen, P., Blumberg, W. E.,* New York: Academic Press 1966.
9. *Pickett, S. M., Riggs, A. F., Larimer, J. L.:* Science *151*, 1005 (1966).
10. *Burns, R. C., Holsten, R. D., Hardy, R. W. F.:* Dupont Innovation *2*, 5 (1970).
11. *Boardman, N. K.:* Advan. Enzymol. *30*, 1 (1968).
12. *Cheniae, G. M., Martin, I. F.:* Plant Physiol. *44*, 351 (1969).
13. *Ginsberg, A. P., Martin. R. L., Sherwood, R. C.:* Inorg. Chem. *7*, 932 (1968).
14. *Bleaney, B., Bowers, K. D.:* Proc. Roy. Soc. (London) (A) *214*, 451 (1952).
15. *Hansen, A. E., Ballhausen, C. J.:* Trans. Faraday Soc. *61*, 631 (1965).
16. *Kramers, H. A.:* Physica *1*, 182 (1934).
17. *Zener, C.:* Phys. Rev. *82*, 403 (1951).
18. *Anderson, P. W.:* Phys. Rev. *115*, 2 (1959).
19. *Van Vleck, J. H.:* The Theory of Electric and Magnetic Susceptibilities. Oxford: University Press 1932.
20. *Griffith, J. S.:* The Theory of Transition-metal ions. Cambridge: University Press 1971.
21. *Kambe, K.:* J. Phys. Soc. Japan *5*, 48 (1950).
22. *Gibson, J. F., Hall, D. O., Thornley, J. H. M., Whatley, F. R.:* Proc. Natl. Acad. Sci. *56*, 987 (1966).
23. *Condon, E. U., Shortley, G. H.:* The Theory of Atomic Spectra. Cambridge: University Press 1953.
24. *Thornley, J. H. M., Gibson, J. F., Whatley, F. R., Hall, D. O.:* Biochem. Biophys. Res. Commun. *24*, 877 (1966).
25. *Moleski, C., Moss, T. H., Orme-Johnson, W. H., Tsibris, J. C. M.:* Biochim. Biophys. Acta *214*, 548 (1970).
26. *Kimura, T., Tasaki, A., Watari, H.:* J. Biol. Chem. *245*, 4450 (1970).
27. *Beinert, H., Palmer, G.:* Advan. Enzym. *27*, 105 (1965).
28. *Title, R. S.:* Phys. Rev. *131*, 623 (1963).
29. *Fano, U., Racah, G.:* Irreducible tensorial sets. New York: Academic Press 1959.
30. *Wigner, E. P.:* Group theory and its application to the quantum mechanics of atomic spectra. New York: Academic Press 1959.

31. *Harris, E. A., Owen, J.:* Phys. Rev. Letters *11*, 9 (1963).
32. *Huang, N. L., Orbach, R.:* Proc. of the int. conf. on magnetism. Inst. of Physics and Phys. Soc. 3 (1964).
33. *Edmonds, A. R.:* Angular momentum in quantum mechanics. Princeton: University Press 1957.
34. *Aitken, A. C.:* Statistical mathematics. Oliver and Boyd 1949.
35. *Howell, K. M.:* Revised tables of the 6 j symbols. Research report US 59-1, University of Southampton (1959).
36. *Uryû, N., Friedberg, S.:* Phys. Rev. *140*A, 1803 (1965).
37. *Weyl, H.:* The Classical Groups. Princeton: University Press 1946.
38. *Griffith, J. S.:* Mol. Phys. *3*, 457 (1960).
39. — The irreducible tensor method for molecular symmetry groups. Prentice Hall 1962.
40. *Figgis, B. N.:* Trans. Faraday Soc. *56*, 1553 (1960).
41. *Mc Kim, F. R.:* Proc. Roy. Soc. (London) *A 262*, 281 (1961).
42. *Griffith, J. S.:* Biopolymers, (London) Symposia No. *1*, 35 (1964).
43. *Kotani, M.:* Prog. Theor. Phys., Suppl. No. 17, 4 (1961).

Received August 3, 1971

Thermochemistry of the Chemical Bond

V. Gutmann and U. Mayer

Institut für Anorganische Chemie der Technischen Hochschule, Wien, Austria

Table of Contents

1. First Pauling Postulate (of the Geometric Mean)

According to *Pauling*, (1) the bond energy D_{A-B}^N of a hypothetical purely covalent single bond ("normal covalent bond") A—B is equal to the geometric mean of the bond energies of the corresponding homoatomic single bonds A—A and B—B:

$$D_{A-B}^N = \sqrt{D_{A-A} \cdot D_{B-B}} \tag{1}$$

The postulate of the geometric mean has been derived from some quantum-mechanical calculations and from the empirical fact that the differences Δ' between the actual bond energy D_{A-B} of a single bond A—B and the energy D_{A-B}^N of the corresponding hypothetical purely covalent single bond,

$$\Delta' = D_{A-B} - D_{A-B}^N = D_{A-B} - \sqrt{D_{A-A} \cdot D_{B-B}} \tag{2}$$

may be approximately described by constants x_A and x_B characteristic of the atoms A and B respectively:

$$\Delta' = D_{A-B} - D_{A-B}^N \approx 30\,(x_A - x_B)^2 \tag{3}$$

The x values are known as the electronegativities.

Δ' represents the energy by which the actual bond A—B is more stable than the hypothetical purely covalent bond. According to *Pauling*, this stabilization is due to "resonance" between the hypothetical covalent and ionic "structures". The resonance stabilization energy Δ' increases with increasing polarity of the actual bond

$$\overset{\delta+}{A}-\overset{\delta-}{B}$$

2. Calculation of Δ' (Second Pauling Postulate)

Calculation of Δ' requires the bond energy D_{A-B} of the actual single bond and the bond energies D_{A-A} and D_{B-B} of the homoatomic single bonds A—A and B—B.

For a diatomic molecule, D_{A-B} is equal to the dissociation energy into A and B in the gas phase provided that A—B represents a single bond. For a polyatomic binary molecule AB_n with n single bonds of the same kind, the bond energy D_{A-B} is obtained by dividing the dissociation energy into the gaseous atoms by n. In molecules containing two or more different bonds, the bond energy of one bond may be calculated, if the bond energies of the remaining bonds are known from other molecules. This procedure involves the assumption that the D_{A-B} terms are approximately constant for different molecules.

For elements which form diatomic molecules A_2 with single bonds, D_{A-A} is equal to the energy of dissociation. This applies to H_2, the halogens, the alkali metals, silver and copper. A few other elements form homoatomic molecules A_m. In the P_4—As_4—and Sb_4-molecules, for instance, each element is bonded to three other atoms, as in compounds of type AB_3. As there are six A—A bonds in the A_4-molecules, the bond energy D_{A-A} is one sixth of the energy of dissociation for the reaction

$$A_{4(g)} \rightarrow 4\,A_{(g)}.$$

Strictly speaking, this D_{A-A} value should be corrected for steric strain because the hybridization of A in the A_4-molecules (bond angle $60°$) does not correspond to that of AB_3 compounds. Furthermore, the

D_{A-A} value obtained in this way can be used only in calculations of Δ'-values for AB_3 compounds (trivalent state) and not, for instance, for AB_5 molecules (pentavalent state). In an analogous manner the energy of dissociation of the S_8- and Se_6-molecules may be used to obtain the bond energies D_{S-S} and D_{Se-Se} respectively. The bond energies D_{A-A} for group IV-elements can be calculated from the energies of atomization of the modification which crystallizes in the diamond lattice. The bond energy is equal to half the values of the energy of atomization. For AB_4 compounds of these elements, the hybridization is the same as in the diamond lattice.

For all other elements, the homoatomic single bond energies cannot be obtained directly from thermochemical data. In order to obtain such bond energies, *Pauling* has made the following assumption which will be referred to as the second *Pauling* postulate.

The enthalpy of formation ΔH_f of a compound in standard state from the elements in the standard state is considered equal to the heat of formation $\overline{\Delta H}_f$ of the gaseous compound from the gaseous (usually hypothetical) element molecules. With this assumption (see below) the standard enthalpy of formation of a compound is equal to the sum of the Δ values of the bonds occuring in the compound:

$$- \Delta H_f = \Sigma \Delta \tag{4}$$

$$\Delta = D_{A-B} - \tfrac{1}{2}\,(D_{A-A} - D_{B-B}) \tag{5}$$

It should be noted that Δ is the difference between D_{A-B} and the arithmetic mean of the homoatomic bond energies.

The second postulate requires that the bond energy between two atoms of elements in their standard states be equal to the energy of the homoatomic single bond of the gaseous (in many cases only hypothetically existent) molecules. Deviations from this postulate if standard states are condensed phases are considered to be due to *van der Waals* interactions. Furthermore, it is required that the resulting compound contain no multiple bonds and that the bond energies for the standard state be about the same as for the gaseous molecule. *Van der Waals* energy terms for the compound and the elements all in the standard state should cancel out.

Eq. (4) must be modified for compounds of nitrogen or (and) oxygen because these elements contain multiple bonds in their standard state (see below).

The electronegativity values are calculated according to *Pauling* by use of the equation

$$\Delta = 23\,(x_A - x_B)^2 \tag{6}$$

3. Ionic Character of a Bond According to Pauling

According to *Pauling*, Δ' for a bond A—B is due to the polarity of the bond. Since Δ'-values can be approximately calculated from the electronegativities of the atoms forming the bond, the polarity of the bond seems to be related to the differences in electronegativities of A and B. The ionic character I_P of a bond A—B in % may be calculated according to *Pauling* by the following "empirical" formula:

$$I_P = (1 - e^{-1/4(x_A - x_B)^2}) \cdot 100 \qquad (7)$$

Thus the ionic character of a bond between atoms of the same electronegativity is zero. 100% ionic character is expected only for electronegativity differences of infinite value. For this bond Δ' should also approach infinity, e. g. the energy of dissociation should be infinitely high.

The factor $1/4$ in the exponent was chosen in order to obtain agreement with the ionic character for HCl, HBr and HI as estimated from the dipole moments.

Unfortunately there is no unambiguous experimental method available for measuring the polarity of a bond and hence Eq. (7) cannot be tested[1].

4. Applicability and Limitations of the Second Pauling Postulate

The second Pauling postulate allows the calculation of homoatomic single bond energies for those elements for which thermochemical data are not directly available. By calculating Δ from formula (4) for a bond A—B in a molecule AB_n with n equal bonds, D_{A-A} is obtained from formula (5) if D_{B-B} (B for example halogen) is known; using Eq. (2) the Δ'_P-value is obtained. To test the applicability of the second postulate, Δ'-values calculated from known homoatomic bond energies are compared with Δ'_P-values calculated using the second postulate. The results are listed in Table 1. Δ_P denotes the value calculated from Eq. (4), D_{A-A}^P the bond energy obtained from Δ_P and Eq. (5). In these calculations the bond energies D_{B-B} of the halogens were assumed to be known except for compounds of hydrogen where the value D_{H-H} was used; for interhalogen compounds the bond energy of the respective heavier element was used.

---

[1] Values for the ionic character of HCl from nuclear quadrupole measurements[2] are up to 45% while calculation from dipole moment and internuclear distance gives 17%[1].

Table 1

	ΔH_f	Δ_P	D_{A-A}^P	Δ_P'	Δ'	Δ	D_{A-B}
LiF	−146.3	146.3	neg.		106.7	106.2	137
LiCl	− 97.7	97.7	neg.		76.8	73.5	115
LiBr	− 83.7	83.7	neg.		67.0	65.5	101
LiI	− 64.8	64.8	neg.		50.9	50.5	81
NaF	−136.0	136	neg.		81.8	80.0	107
NaCl	− 98.2	98.2	neg.		66.3	60.4	98
NaBr	− 86.0	86.0	neg.		59.7	56.3	88
NaI	− 68.8	68.8	neg.		46.0	44.3	71
KF	−134.5	134.5	neg.		97.2	93.8	118
KCl	−104.2	104.2	neg.		74.8	66.1	101
KBr	− 93.7	93.7	neg.		67.6	62.1	91
KI	− 78.3	78.3	neg.		56.3	53.0	77
RbF	−131.3	131.3	neg.		99.1	95.3	119
RbCl	−102.9	102.9	neg.		76.9	67.6	102
RbBr	− 93.0	93.0	neg.		67.6	61.5	90
RbI	− 78.5	78.5	neg.		57.2	53.5	77
CsF	−126.9	126.9	neg.		101.5	97.5	121
CsCl	−103.5	103.5	neg.		76.4	66.8	101
CsBr	− 94.3	94.3	neg.		69.1	62.8	91
CsI	− 80.5	80.5	neg.		55.6	51.8	75
CuCl	− 32.2	32.2	53.6	32.2	35.0		88
CuBr	− 25.1	25.1	59.7	25.2	30.7		78
AgCl	− 30.3	30.4	25.2	33.8	24.3		72
AgBr	− 23.8	23.8	44.3	23.8	26.4		69
AgI	− 14.9	14.9	54.1	15.8	22.4		60
HF	− 64.2	64.2	37.0	72.6	72.6		134.8
HCl	− 22.1	22.1	58.0	25.5	25.5		103.2
HBr	− 8.7	8.7	53.4	12.9	18.2		87.5
HI	+ 6.2	−6.2	51.0	−1.5	10.1		71.4
H_2O	− 68.3	47.2	22.6	62.1	51.7		110.6
H_2S	− 4.8	2.4	56.8	6.0	8.0		82.9
H_2Se	+ 20.5	−10.2	49.2	−5.0	−0.8		66.5
NH_3	− 11.0	22.1	38.4	30.1	30.1		93.4
PH_3	+ 2.2	−0.7	50.4	4.2	6.0		76.6
AsH_3	+ 41.0	−13.7	46.0	−7.7	−3.4		61.4
BrF	− 14.0	14.0	53.4	14.8	17.8		59.2
ClF	− 13.3	13.3	58.0	14.5	14.5		60.8
ICl					4.5		50.3
IBr					1.7		42.5
NF_3	− 29.7	28.4	38.4	27.3	27.3		65.0
NCl_3					−1.1		46.0
PF_3	−218.3	72.8	50.6	73.3	74.4		116.4
PCl_3	− 73.9	24.6	45.4	25.0	23.5		76.3
PBr_3	− 41.5	13.8	51.3	13.8	15.4		62.5
PI_3	− 10.9	3.6	46.7	2.9	2.3		44

131

Table 1 (continued)

	ΔH_f	Δ_P	D_{A-A}^P	Δ'_P	Δ'	Δ	D_{A-B}
AsF$_3$	−226.8	75.6	40.8	75.5	75.8		114.3
AsCl$_3$	− 80.2	26.7	40.2	27.4	27.4		75.8
AsBr$_3$	− 46.6	15.5	44.7	15.5	17.7		60.9
AsI$_3$	− 13.6	4.5	46.3	4.8	7.5		45.7
SbF$_3$	−217.2	72.4	29.8	72.6	71.1		105.6
SbCl$_3$	− 91.3	30.4	29.8	32.5	31.0		74.4
SbBr$_3$	− 62.1	20.7	37.7	20.9	23.9		62.6
SbI$_3$	− 23.0	7.7	43.5	7.9	13.2		47.5
BiF$_3$	− 211	70.3	10.6	74.2			93.9
BiCl$_3$	− 90.6	30.2	13.6	37.9			66.0
BiBr$_3$	− 63.0	21.0	20.7	23.5			54.4
CF$_4$	−219	54.8	85.8	59.9	60.0		115.9
CCl$_4$	− 33.3	8.3	81.6	9.2	7.6		78.1
CBr$_4$	− 0.9	0.2	86.3	3.2	3.7		66.4
SiF$_4$	−386	96.5	52.8	97.2	97.4		141.2
SiCl$_4$	−153	38.2	49.0	38.3	36.5		91.7
SiBr$_4$	− 95.1	23.8	55.3	23.9	25.3		74.5
SiI$_4$	− 31.6	7.9	60.3	9.3	12.5		56.1
GeCl$_4$	−129.4	32.3	40.8	33.0	30.9		81.7
GeBr$_4$	− 78.5	19.6	46.9	19.6	20.8		66.1
GeI$_4$	− 30.5	7.6	50.3	8.1	10.6		50.8
SnCl$_4$	−130.3	32.6	31.8	34.5	32.0		77.5
SnBr$_4$	− 97.1	24.3	36.7	24.5	25.1		65.7
BF$_3$	−270.1	90.0	94.0	96.6			155.5
BCl$_3$	− 96.8	32.3	94.0	34.5			108.3
BBr$_3$	− 57.5	19.2	96.3	23.9			90.4
AlF$_3$	−360.8	120.3	4.2	127.9			140.9
AlCl$_3$	−168.6	56.2	33.0	58.0			101.7
AlBr$_3$	−125.8	41.9	43.7	41.8			86.8
AlI$_3$	− 75.2	25.1	50.9	25.2			68.6
GaCl$_3$	−125	41.7	28.8	44.3			85.1
GaBr$_3$	− 92.4	30.8	37.9	31.0			72.8
GaI$_3$	− 51.0	17.0	43.5	17.2			56.8

Values of dissociation energies were taken with few exceptions from *T. L. Cottrell*, "The Strength of Chemical Bonds", Butterworth, London (1958). Standard enthalpies of formation, heats of fusion and vaporization or heats of sublimation were taken from *F. D. Rossini, D. D. Wagman, W. H. Evans, S. Levine* and *I. Jaffee*, "Selected Values of Chemical Thermodynamic Properties", National Bureau of Standards Circular 500 (1952), except for cases where more recent data have been published since. Improved enthalpy data for a number of fluorides and chlorides have been compiled by *H. A. Skinner*, "Pure and Applied Chemistry 8, 113 (1964). Heats of atomization of the elements were taken from *D. R. Stull* and *G. C. Sinke*, "Thermodynamic Properties of the Elements", Advances in Chemistry Series No. 18, American Chemical Society, Washington (1956). Enthalpy data for GeBr$_4$ and

GeI$_4$, not given in NBSC 500, have been reported by *D. E. Evans* and *R. E. Richards*, J. Chem. Soc. *1952*, 1292. The standard enthalpy of formation of AlF$_3$ has recently been redetermined by *E. Rudzitis*, *H. M. Feder* and *W. N. Hubbard*, Inorg. Chem. *6*, 1716 (1967); the heat of sublimation of AlF$_3$ is that given in JANAF Thermochemical Tables quoted in *Kirk-Othmer*, "Encyclopedia of Chemical Technology", second ed., vol. 9, 529 (1966). Heats of vaporization for Al- and Ga-halides (except AlF$_3$) were recalculated from the original literature: *W. Fischer* and *O. Rahlfs*, Z. anorg. Chem. *205*, 1 (1932) and *W. Fischer* and *O. Jübermann*, Z. anorg. Chem. *227*, 227 (1936); the corresponding heats of fusion were estimated according to *O. Kubaschewski* and *E. Ll. Evans*, "Metallurgische Thermochemie", Verlag Technik, Berlin (1959). Thermochemical data for Bi- halides are from Gmelins Handbuch der anorganischen Chemie, Wismut, Ergänzungsband 1964, Verlag Chemie, Weinheim/Bergstraße. Data for BrF are from *W. H. Evans*, *T. R. Munson*, *D. D. Wagman*, J. Res. nat. Bur. Stand. *55*, 147 (1955) and *L. Stein* in "Halogen Chemistry" (ed. *V. Gutmann*), vol. 1, p. 154, Academic Press, London—New York (1967).

Table 1 reveals that, for the halides of group IV and group V elements, agreement between Δ' and Δ'_P-values is reasonably good; it is least satisfactory for the iodides, which have low enthalpies of formation. Agreement between Δ' and Δ'_P is perfect for HF as well as for NF$_3$, NH$_3$, HCl and ClF, despite low enthalpies of formation because formation of these compounds from the elements under standard conditions proceeds as a gas phase reaction.

For the other hydrogen compounds listed in Table 1 which have low or positive enthalpies of formation, Δ' and Δ'_P values disagree by up to more than 100 % [2]).

The Δ'-values for AsH$_3$ and H$_2$Se are negative. This may be due partly to inaccurate ΔH_f values and partly to deviations from the first Pauling postulate.

Table 1 shows that the postulate is not applicable to alkali metal halides MX. By use of D$_{X-X}$, negative values for the M—M—bond energies are obtained from Δ_P; the differences between Δ_P and Δ-values may be up to 60 %.

We shall now proceed to discuss the differences between Δ'_P and Δ'-values using Born-Haber cycles.

Let us consider a halide AB$_n$ with n equal bonds A—B. Cycle, Fig. 1. St denotes the state of aggregation under standard conditions (298 °K, 1 at), ΔH_f the standard enthalpy of formation for AB$_n$, $\overline{\Delta H}_f$ the enthalpy of formation of the gaseous compound AB$_n$ from the gaseous element molecules A$_m$ and B$_2$. ΔH_f^{At} represents the atomic enthalpy of formation of the gaseous compound from the atoms in the gas

---

[2]) The Δ'_P-values have been obtained by using D$_{H-H}$ = 104.2 kcal. Better agreement is observed when use is made of D$_{I-I}$ = 36.1 kcal, D$_{Br-Br}$ = 46.1 kcal and D$_{O-O}$ = 33.2 kcal for which Δ'_P becomes 5.8 for HI, 15.9 for HBr and 54.9 for H$_2$O.

phase (this quantity is numerically equal to the number of bonds n multiplied by the bond energy D_{A-B}). A_m denotes a (usually hypothetical) gaseous element molecule consisting of m atoms, each of them having the same number of valencies as A in AB_n; furthermore, the hybridization of A in A_m and AB_n should be the same. For compounds AB_3, for instance, with trivalent A, the simplest case of a molecule A_m is that with $m = 4$, as is known for phosphorous, arsenic and antimony.

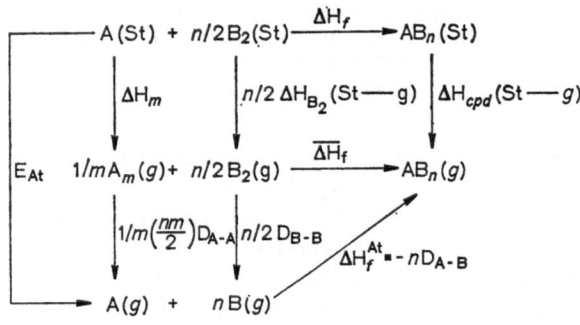

Fig. 1

For a molecule A_m in which each atom has n valencies, $(n/2)$ m bonds of energy D_{A-A} must be broken in order to obtain the atoms A. ΔH_m is the mth fraction of the energy required to transfer m gram-atoms A from the standard state into the gaseous molecule A_m. E_{At} is the energy of atomization of the element A. According to the cycle shown in Fig. 1, we obtain

$$\frac{1}{m}\left(\frac{n}{2}m\right) D_{A-A} + \frac{n}{2} D_{B-B} + \Delta H_f^{At} = \overline{\Delta H}_f \qquad (8)$$

$$\Delta H_f^{At} \equiv -n\, D_{A-B} \qquad (9)$$

$$-\overline{\Delta H}_f = n\left[D_{A-B} - \tfrac{1}{2}(D_{A-A} + D_{B-B})\right] = n \cdot \Delta \qquad (10)$$

When $\overline{\Delta H}_f$ is known, Δ may be calculated from (10) and with the value of D_{B-B} (which is assumed to be known), Δ' is also obtained. $\overline{\Delta H}_f$ is related to ΔH_f by Eq. (11), in which cpd denotes the compound under consideration:

$$\overline{\Delta H}_f = \Delta H_f + \Delta H_{cpd}(\text{St} \to \text{g}) - \frac{n}{2}\, \Delta H_{B_2}(\text{St} \to \text{g}) - \Delta H_m \qquad (11)$$

Unfortunately, ΔH_m is unknown for most of the elements. According to the second postulate, $\Delta H_f \equiv \overline{\Delta H}_f$, compensation of the rest of the energy quantities of cycle (11) is assumed according to

$$\Delta H_m + \frac{n}{2}\, \Delta H_{B_2}\,(St \rightarrow g) = \Delta H_{cpd}\,(St \rightarrow g) \tag{12}$$

Table 2 shows that the differences between $\Delta H_{cpd}\,(St \rightarrow g)$ and $\Delta H_m + \frac{n}{2}\Delta H_{B_2}(St \rightarrow g)$ are very small indeed for AsF_3 and $AsCl_3$; Δ' and Δ'_P are therefore almost identical (Table 1). There are differences for $AsBr_3$ (6.9 kcal/mol) and AsI_3 (6.8 kcal/mol). The enthalpy of formation is considerably smaller for AsI_3 than for $AsBr_3$ (Table 1) and hence Δ'_P deviates from Δ' by 36% for AsI_3 and by 13% for $AsBr_3$.

Table 2

	AsF_3	$AsCl_3$	$AsBr_3$	AsI_3
$\Delta H_m\,[As\,(sol) \rightarrow 1/4\,As\,(g)]$	8.6	8.6	8.6	8.6
$\frac{3}{2}\Delta H_{B_2}\,(St \rightarrow g)$	0.0	0.0	11.1	22.2
$\Delta H_{cpd}\,(St \rightarrow g)$	8.5	8.7	12.8	22.0
$\Delta H_m + \frac{3}{2}\Delta H_{B_2}\,(St \rightarrow g)$	8.6	8.6	19.7	28.8

For HF and HCl, Eq. (13) is satisfied (Table 3) because their formation under standard conditions is a gas phase reaction; Δ' and Δ'_P are therefore identical.

Table 3

	HF	HCl	HBr	HI
$\Delta H_m\,[1/2\,H_2(g) \rightarrow 1/2\,H_2(g)]$	0.0	0.0	0.0	0.0
$1/2\,\Delta H_{B_2}(St \rightarrow g)$	0.0	0.0	3.7	7.4
$\Delta H_{cpd}(St \rightarrow g)$	0.0	0.0	0.0	0.0
$\Delta H_m + 1/2\,\Delta H_{B_2}(St \rightarrow g)$	0.0	0.0	3.7	7.4

The differences between the left and the right sides of Eq. (12) are small for HBr (3.7 kcal/mol) and HI (7.4 kcal/mol), but the small enthalpy of formation for the latter compound makes the deviation between Δ'_P and Δ' considerable, namely 115% using D_{H-H} and 42% using D_{I-I}.

135

The situation is similar for H_2S and H_2Se (Table 4).

Table 4

	H_2S		H_2Se
ΔH_m [S(sol) → $^1/_8$ $S_8(g)$]	3.05	ΔH_m [Se(sol) → $^1/_6$ $Se_6(g)$]	5.9
ΔH_{B_2}(St → g)	0.0		0.0
ΔH_{cpd}(St → g)	0.0		0.0
$\Delta H_m + \Delta H_{B_2}$(St → g)	3.05		5.9

Eq. (12) is inapplicable to the alkali metal halides (Table 5) with differences between 20 and 60 kcal; accordingly the differences between Δ and Δ_P values are also high (Table 1), although these compounds have rather high enthalpies of formation.

Table 5

	NaF	NaCl	NaBr	NaI
ΔH_m [Na(sol) → $^1/_2$ $Na_2(g)$]	16.9	16.9	16.9	16.9
$^1/_2 \Delta H_{B_2}$(St → g)	0.0	0.0	3.7	7.4
ΔH_{cpd}(St → g)	73.5	54.7	49.7	47.9
$\Delta H_m + ^1/_2 \Delta H_{B_2}$(St → g)	16.9	16.9	20.6	24.3

Since alkali metal halides have high heats of sublimation, the term on the right side of Eq. (12) is larger than that on the left side. Thus, according to Eq. (11), $|\Delta H_f|$ has a higher value than $|\overline{\Delta H}_f|$ and Δ_P values calculated from ΔH_f are too high.

The situation is different for $AsBr_3$, AsI_3, HBr, HI, H_2S and H_2Se, where Δ_P values are too small as compared to the true values Δ'.

Eq. (4) has to be modified for compounds containing nitrogen and oxygen. These elements are gases in their standard state but contain multiple bonds. Eq. (4) is applicable only to single bonds, therefore ΔH_f must be corrected by the energy required to change the multiple bond into the single bond. For the homoatomic single bond energies, the bond energies N—N in hydrazine and O—O in hydrogen peroxide have been accepted (7):

$$\frac{1}{2} N_{2(g)} \xrightarrow{\Delta H_m^N} \frac{1}{m} N_{m(g)} \xrightarrow{\frac{1}{m}(\frac{2}{2}m) D_{N-N}} N_{(g)}$$
$$\underbrace{\hspace{6cm}}_{E_{At}^N}$$

$$\frac{1}{2} O_{2(g)} \xrightarrow{\Delta H_m^O} \frac{1}{m} O_{m(g)} \xrightarrow{\frac{1}{m}(\frac{2}{2}m) D_{O-O}} O_{(g)}$$
$$\underbrace{\hspace{6cm}}_{E_{At}^O}$$

The correction terms ΔH_m^N and ΔH_m^O are calculated using $E_{At}^N = 113$ kcal, $E_{At}^O = 59.2$ kcal, $D_{N-N} = 38.4$ kcal and $D_{O-O} = 33.2$ kcal as $\Delta H_m^N = 55.4$ kcal (trivalent, sp³-hybridized nitrogen) and $\Delta H_m^O = 26.0$ (divalent, sp³-hybridized oxygen). Eq. (4) then becomes for compounds containing n_N nitrogen atoms:

$$- \Delta H_{f(\text{corr.})} = \Delta H_f - n_N \cdot 55 \cdot 4 = \Sigma \Delta \qquad (4a)$$

and for compounds containing n_O oxygen atoms:

$$- \Delta H_{f(\text{corr.})} = \Delta H_f - n_O \cdot 26 \cdot 0 = \Sigma \Delta \qquad (4b)$$

while formula [11] is reduced to

$$\frac{n}{2} \Delta H_{B_2}(St \to g) = \Delta H_{cpd}(St \to g) \qquad (11a)$$

because ΔH_m^N and ΔH_m^O have been taken into account in $\Delta H_{f(\text{corr.})}$.

For ammonia $\Delta H_{B_2}(St \to g) = 0$ as well as $\Delta H_{cpd}(St \to g)$, since both H_2 and NH_3 are gases under standard conditions. Eq. (11a) is valid in this case and hence there is exact agreement between Δ' and Δ'_p. For water $\Delta H_{B_2}(St \to g) = 0$ and $\Delta H_{cpd}(St \to g) = 10.5$ kcal/mol (heat of vaporization). Thus Eq. (11a) is not valid and $|\Delta H_{f(\text{corr.})}|$ is according to (11) by 10.5 kcal higher than $|\overline{\Delta H_f}|$ and hence Δ'_p is higher than the true value Δ'.

Formulas (11) and (11a) can be expanded for compounds formed from more than two different elements. For compounds containing both nitrogen and oxygen, one obtains

$$- \Delta H_{f(\text{corr.})} = \Delta H_f - n_N \cdot 55 \cdot 4 - n_O \cdot 26 \cdot 0 = \Sigma \Delta \qquad (4c)$$

There is no general criterion for the applicability of the second postulate to compounds AB_n for which only one of the homoatomic single bond energies D_{A-A} and D_{B-B} is available. A necessary but not sufficient condition is that, by variation of B in AB_n (such as from fluoride to iodide), the bond energy D_{A-A} calculated from (4) and (5) should remain approximately constant and thus the ΔH_m-values calculated from (13) (see energy cycle) should also be constant:

$$\Delta H_m + \frac{n}{2} D_{A-A} = E_{At} \tag{13}$$

However, as can be seen from Eq. (12), $\frac{n}{2} \Delta H_{B_2}(St \rightarrow g)$ and ΔH_{cpd} $(St \rightarrow g)$ may vary in such a way, that nearly constant "ΔH_m" and "D_{A-A}" values are obtained although they may differ more or less from the true values.

This applies for instance to the boron (III)-halides for which nearly the same D_{B-B} values are obtained (Table 1). With $\Delta H_f \equiv \overline{\Delta H}_f$ assumed to be valid, Eq. (12) is applicable. For BF_3 and BCl_3 with $\Delta H_{B_2}(St \rightarrow g) = 0$ and $\Delta H_{cpd}(St \rightarrow g) = 0$, the result is $\Delta H_m = 0$, and for BBr_3 with $\frac{3}{2} \Delta H_{B_2}$ $(St \rightarrow g) = 11.1$ and $\Delta H_{cpd}(St \rightarrow g) = 7.3$, one obtains $\Delta H_m = -3.8$. This negative value for ΔH_m appears unreasonable. Values $\Delta H_m = 0$ *do* exist in case of the diamond modification of C, Si, Ge and Sn for the tetravalent sp^3-hybridized state of these elements. For sp^2-hybridized boron, $\Delta H_m = 0$ is improbable in view of the complicated structures of crystalline boron.

The application of the second postulate to aluminium halides leads to vastly different ΔH_m- and hence to different D_{Al-Al} values (Table 6 and Table 1).

Table 6

	AlF_3	$AlCl_3$	$AlBr_3$	AlI_3
$\Delta H_{cpd}(St \rightarrow g)$	71.2	28.0	23.0	23.3
$\frac{3}{2} \Delta H_{B_2}(St \rightarrow g)$	0.0	0.0	11.1	22.2
$\Delta H_m[Al(sol) \rightarrow \frac{1}{m} Al_m(g)]$	71.2	28.0	11.9	1.1

The value $\Delta H_m = 71.2$ for AlF_3 is doubtless too high; from the heat of atomization of aluminium ($E_{At} = 77.5$ kcal) and Eq. (13) ($n = 3$), a value D_{Al-Al} of 4.2 kcal is obtained, which appears far too low. Thus, according to Eq. (11), $\overline{\Delta H}_f$ is numerically smaller than ΔH_f and therefore values Δ_P and Δ'_P calculated from ΔH_f are too high compared with the true values Δ and Δ'. On the other hand, ΔH_m for AlI_3 is doubtless too low, consequently Δ_P and Δ' calculated according to the second postulate must also be too low. With a reasonable assumption of $\Delta H_m \approx 20$ kcal (compare the value given for boron) corresponding to $D_{Al-Al} = 38,3$ kcal, figures are obtained which are presented in Table 7.

Table 7

		AlF_3	$AlCl_3$	$AlBr_3$	AlI_3
Δ'		103.2	54.6	44.9	31.5
$Q_{mol} = \dfrac{\Delta'}{D_{Al-X}} \cdot 100$		73.4	53.6	51.8	45.9
Q_a		53.2	32.3	30.9	25.9

It can be seen from Table 1 and Table 7 that Δ'_P values calculated according to the second postulate are higher by 24 % for AlF_3 and lower by 20 % for AlI_3 as compared with the actual values Δ'. There is reasonable agreement of Δ'_P and Δ'-values for $AlCl_3$ and $AlBr_3$.

Various values for ΔH_m are also obtained by applying the second postulate to $GaCl_3$ (21.8 kcal), $GaBr_3$ (8.3 kcal) and GaI_3 (-0.3 kcal). Application of the second postulate to gallium halides is obviously not correct: the value of -0.3 kcal is certainly too small and hence the same applies to Δ'_P.

As stated above, the true value for the boron-boron bond D_{B-B} is not directly available from thermochemical data. However, in B_2X_4 halides a boron-boron-bond is present. Boron shows the same valency and nearly the same hybridization in B_2X_4 halides as in BX_3 halides. Assuming that the B—X bonds are of equal strength in BX_3 and B_2X_4, a value of the B—B bond energy of 72.4 kcal for B_2F_4 and 79.0 kcal for B_2Cl_4 is calculated from thermochemical data (3). Values of D_{B-B} obtained by this method should correspond to the true value of the homoatomic sp^2-hybridized B—B bond energy if the B—X bonds in B_2X_4 are nearly nonpolar. The value of 79.0 kcal therefore seems to be the best approximation available. With $D_{B-B} = 79.0$ kcal, ΔH_m becomes 22.5 kcal (compared with $\Delta H_m = 0$ using the second postulate); figures obtained with these values are listed in Table 8.

Table 8

		BF_3	BCl_3	BBr_3
Δ'		101.5	40.7	30.1
$Q_{mol} = \dfrac{\Delta'}{D_{B-X}} \cdot 100$		65.3	37.6	33.3
Q_a		43.8	19.9	17.1

139

For boron (III) halides according to (11) $\overline{\Delta H}_f$ is numerically higher than ΔH_f and Δ'_P therefore smaller than Δ', namely by 5 % for BF$_3$, by 15 % for BCl$_3$ and by 21 % for BBr$_3$ (the influence of back donation will be discussed later).

In some cases it is possible to deduce homoatomic bond energies by extrapolation from the trend of bond energies observed within the periodic table. Figures are given in Table 9.

Table 9

	D_{E-E} kcal	$E_{At}[E(\text{sol}) \rightarrow E(g)]$ kcal	$\Delta H_m[E(\text{sol}) \rightarrow \frac{1}{m} E_m(g)]$ kcal
Li	26.4	38.4	25.2
Na	18.0	25.9	16.9
K	12.2	21.4	15.3
Rb	11.6	19.6	13.8
Cs	10.8	18.7	13.3
Ag	39.2	68.0	48.4
Cu	48.4	81.0	56.8
S	53.8	56.9	3.1
Se	43.5	49.4	5.9
P [a] (white)	48.0	75.6	3.6
P [a] (red)	48.0	79.8	7.8
As [a]	40.3	69.0	8.6
Sb [a]	32.4	60.8	12.2
C (diamond)	85.2	170.4	0.0
Si	52.5	105.0	0.0
Ge	44.5	89	0.0
Sn (grey)	35.7	71.4	0.0
F$_2$	37.0		
Cl$_2$	58.0		
Br$_2$	46.1		
J$_2$	36.1		
H$_2$	104.2		

[a] D_{E-E} values not corrected for steric strain in the A$_4$ molecules. The bond energies of the "sp^3" hybridized bonds will be a few percent higher.

Thus, D_{Bi-Bi} can be extrapolated (Table 9) to be ≈ 20 kcal, corresponding to $\Delta H_m \approx 20$ kcal and $D_{Te-Te} \approx 35$ kcal corresponding to $\approx \Delta H_m \approx 12$ kcal. Figures for bismuth (III) halides derived by assuming $D_{Bi-Bi} \approx 20$ kcal are listed in Table 10; for BiF$_3$ and BiCl$_3$ the Δ'_P-values are higher than the true values Δ'.

Table 10

	BiF$_3$	BiCl$_3$	BiBr$_3$
Δ'	66.8	31.9	24.0
$Q_{mol} = \dfrac{\Delta'}{D_{Bi-X}} \cdot 100$	71.1	48.3	44.1
Q_a	50.4	28.0	24.6

5. A New Approach to the Evaluation of the Ionic Character of Bonds by Thermochemical Data

a) Diatomic Molecules

The quantity Δ' represents the energy by which the actual polar bond A—B of a diatomic molecule is stabilized due to its partial ionic character compared with the hypothetical purely covalent bond A—B. The quotient Q of Δ' and the total interaction energy D_{A-B} of the atoms A and B multiplied by 100 should therefore be a useful energetic measure for the ionic character of the bond.

$$Q = \frac{\Delta'}{D_{A-B}} \cdot 100 \qquad (14)$$

Figs. 2 and 3 show the energy relationship for two different diatomic molecules A_1-B_1 and B_2-A_2. Conditions were chosen so that the Δ'-

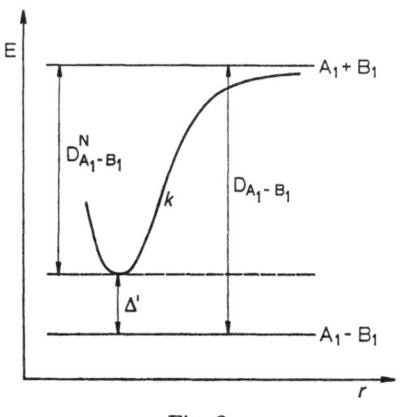

Fig. 2

values are identical for both bonds, while $D_{A_1-B_1}$ is considerably higher than $D_{A_2-B_2}$. k denotes the potential curve for the hypothetical purely covalent bond as a function of the internuclear distance r. For molecule A_1-B_1 the relative deviation of the energy content $D_{A_1-B_1}^N$ of the purely covalent bond from the actual bond energy $D_{A_1-B_1}$ is considerably smaller than for molecule 2, due to its higher value $D_{A_1-B_1}$. It may therefore be concluded that bond 1 is more covalent in character than bond 2 as represented by the corresponding Q-values $Q_1 < Q_2$. In contrast the same ionic character is obtained for both molecules according to the *Pauling* formula (7).

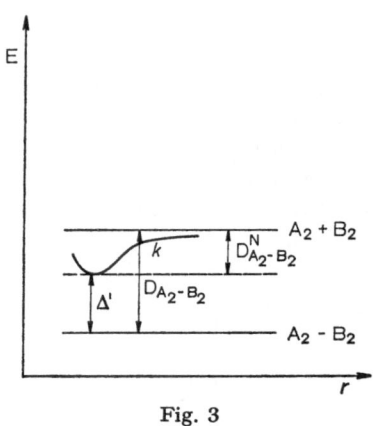

Fig. 3

From Figs. 2 and 3 it can be seen that Δ' approaches zero for a purely covalent bond, which means that Q also approaches zero. With increasing ionic character of the bond, the minimum of the potential curve k becomes less pronounced and finally disappears: Δ' approaches D_{A-B} and quotient Q approaches 100%.

b) Polyatomic Molecules

For a polyatomic molecule AB_n with n equal bonds, the energy of formation $E_{AB_n}^N$ of the corresponding hypothetical purely covalent gaseous molecule from gaseous atoms can be calculated in the same way as for a diatomic molecule. Analogous to Eq. (1) one obtains

$$E_{AB_n}^N = n \cdot D_{A-B}^N = n \cdot \sqrt{D_{A-A} \cdot D_{B-B}} \tag{15}$$

The difference in energy between the actual gaseous molecule AB_n and the hypothetical purely covalent molecule is given by

$$\Delta'_{mol} = |\Delta H_f^{At}| - E_{AB_n}^N \tag{16}$$

The mean bond energy D_{A-B} of a bond $A-B$ in the molecule AB_n is related by definition to ΔH_f^{At} according to formula (10), and hence Δ'_{mol} becomes

$$\Delta'_{mol} = n\left(D_{A-B} - \sqrt{D_{A-A} \cdot D_{B-B}}\right) \tag{17}$$

The ionic character of the molecule is obtained in analogy to formula (14) from the ratio of the stabilization energy Δ'_{mol} to the atomic enthalpy of formation ΔH_f^{At}:

$$Q_{mol} = \frac{\Delta'_{mol}}{|\Delta H_f^{At}|} \cdot 100 = D_{A-B} - \frac{\sqrt{D_{A-A} \cdot D_{B-B}}}{D_{A-B}} \cdot 100 \tag{18}$$

Eq. (18) is formally identical with Eq. (14) for the ionic character of a bond $A-B$ in a diatomic molecule. The difference is, however, that Q_{mol} is not valid for one bond $A-B$ in the molecule AB_n but only for consideration of the whole molecule. Our ideas may be illustrated by considering a molecule AB_3 with three equal bonds

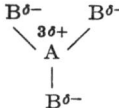

Each atom B carries a negative partial charge of amount $\delta-$ while atom A has the positive partial charge $3\delta+$.

For the evaluation of the polarity or of the ionic character, respectively, for the bond $A-B$ in the molecule AB_3, and for comparison with diatomic molecules, the amount of the negative partial charge at atom B has to be considered. Hence, the polarity or ionic character of the bond $A-B$ in the molecule AB_3 corresponds to that of a bond $>A^{\delta+}-B^{\delta-}$ with the two other bonds being non-polar.

According to *Pauling*, the quantity Δ' for a bond $A-B$ in a diatomic molecule $A^{\delta+}-B^{\delta-}$ is proportional to the polarity of the bond. This can be written as

$$\Delta' = \text{prop.} \frac{\delta^2}{d} \tag{19}$$

if Δ' is interpreted as Coulombic energy term (d is the internuclear distance). Applying this idea to polyatomic molecules AB_n with n equal bonds A—B, one obtains

$$\Delta'_{mol} = \text{prop.} \ \frac{f \cdot \delta^2}{d} \tag{20}$$

δ is the amount of partial charge at atom B and d the internuclear distance A—B. f is a quantity characteristic of the structure of the molecule and the number n of bonds (see below). The stabilization energy Δ'_a for the bond $>A^{\delta+} - B^{\delta-}$ (index a from actual ionic character) is given by

$$\Delta'_a = \text{prop.} \ \frac{\delta^2}{d} \tag{21}$$

and the ionic character Q_{mol} of the molecule according to (18):

$$Q_{mol} = \frac{\Delta'_{mol}}{n \cdot D_{A-B}} \cdot 100 = \text{prop.} \ \frac{\frac{f}{n} \cdot \delta^2}{D_{A-B} \cdot d} \cdot 100 = \text{prop.} \ \frac{f' \cdot \delta^2}{D_{A-B} \cdot d} \cdot 100 \tag{22}$$

with

$$f' = \frac{f}{n} \tag{23}$$

In order to describe the ionic character Q_a of the bond $>A^{\delta+} - B^{\delta-}$, the corresponding bond energy D_a must be known. With D_a the actual ionic character Q_a of the bond A—B in the molecule AB_n is given according to formula (14) by

$$Q_a = \frac{\Delta'_a}{D_a} \ 100 = \text{prop.} \ \frac{\delta^2}{d \cdot D_a} \cdot 100 \tag{24}$$

From the ratio Q_a/Q_{mol} there follows

$$Q_a = Q_{mol} \cdot \frac{D_{A-B}}{f' \cdot D_a} \tag{25}$$

Since D_a is unknown, the following considerations may be helpful: D_a for a purely covalent molecule AB_n must be identical with D_{A-B} since no polarity is present:

$$D_a = D_{A-B} \ \text{(for a purely covalent molecule } AB_n) \tag{26}$$

For a purely ionic bond, both Q_{mol} and Q_a must be 100, which means

$$Q_a = Q_{mol}$$

and with (25):

$$D_a = \frac{D_{A-B}}{f'} \text{ (for a purely ionic molecule } AB_n) \qquad (27)$$

For diatomic molecules, f' equals unity, for polyatomic molecules it is always greater than unity, as will be shown later; hence $D_a \leq D_{A-B}$ for polyatomic molecules. In the case of a partial polar bond $A-B$, D_a may be calculated by making use of the assumption that D_a decreases linearly with increasing ionic character Q_{mol} of the molecule AB_n between the limiting cases of Eqs. (26) and (27):

$$D_a = D_{A-B} - k \cdot D_{A-B} \cdot \frac{Q_{mol}}{100} \qquad (28)$$

constant k must be chosen in such a way that the limiting case (27) is satisfied by Eq. (28). One obtains

$$k = \frac{f'-1}{f'} \qquad (29)$$

$$D_a = D_{A-B} \left[1 - \frac{Q_{mol}}{100} \frac{(f'-1)}{f'} \right] \qquad (30)$$

$$Q_a = Q_{mol} \cdot \frac{100}{100 \cdot f' - Q_{mol}(f'-1)} \qquad (31)$$

From $f' \geq 1$ and $Q_{mol} \leq 100$ it follows that $Q_a \leq Q_{mol}$: thus the actual ionic character Q_a of a bond $A-B$ in a polyatomic molecule AB_n is smaller than the ionic character Q_{mol} of the molecule calculated according to Eqs. (9) and (18). This is a consequence of the accumulation of charge at the central atom. For diatomic molecules Q_a is identical with Q_{mol}.

The calculation of values f' may be demonstrated for a tetrahedral molecule AB_4 (Fig. 4).

According to (20), the stabilization energy Δ'_{mol} is proportional to the potential electrostatic energy V of the (pointlike) charges. V is given

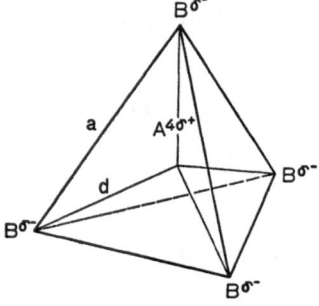

Fig. 4

by the sum of a negative term for the attraction between charges of opposite sign and a positive term for the repulsion between like charges:

$$V = -4 \cdot \frac{4\,\delta^2}{d} + 6 \cdot \frac{d^2}{a}$$

The distance between the negative charges follows from the geometry of the tetrahedron: $a = 1.633\, d$. This leads to

$$|V| = \frac{16\,\delta^2}{d} - \frac{3.67\,\delta^2}{d} = \frac{12.33\,\delta^2}{d}$$

and to $f = 12.33$ and $f' = f/4 = 3.08$.

Table 11. *Calculated f' values for different types of molecules*

Molecule	Structure	angle BAB	Example	f'
AB				1,00
AB$_2$	linear	180°	HgX$_2$	1,75
AB$_2$	angular	104°	H$_2$O	1,68
AB$_2$	angular	92°	H$_2$S,H$_2$Se	1,65
AB$_3$	planar	120°	BX$_3$,AlX$_3$(gas)	2,42
AB$_3$	pyramidal	107°	NH$_3$	2,38
AB$_3$	pyramidal	100°	AsX$_3$	2,35
AB$_3$	pyramidal	94°	PH$_3$	2,32
AB$_4$	tetrahedral	110°	CX$_4$,SiX$_4$	3,08

6. Comparison and Discussion of I_P and Q_a Values

Table 12. *Values for the ionic character of several bonds according to I_P values (calculated from formula (7) using Pauling's electronegativities (1) of the bonded atoms) and Q_a values (calculated by the use of Δ', D_{A-B} and f' values, formula (18) and (31)). Improved Pauling electronegativities have been reported by Allred (4), the resulting differences from I_P-values given in this table are, however, not significant.*

	Q_{mol}	I_P	Q_a
LiF	77.9	89.0	77.9
LiCl	66.9	63.0	66.9
LiBr	66.3	55.0	66.3
LiI	62.8	43.0	62.8
NaF	76.5	90.5	76.5[a]
NaCl	67.6	66.5	67.6
NaBr	67.8	59.0	67.8
NaI	64.8	47.0	64.8
KF	82.4	92.0	82.4
KCl	74.0	70.0	74.0
KBr	74.3	63.0	74.3
KI	73.1	51.0	73.1
RbF	83.2	92.0	83.2
RbCl	75.4	70.0	75.4
RbBr	75.1	63.0	75.1
RbI	74.3	51.0	74.3
CsF	83.9	93.0	83.9
CsCl	75.6	73.0	75.6
CsBr	75.9	66.5	75.9
CsI	74.0	55.0	74.0
CuCl	39.8	26.0	39.8[b]
CuBr	39.4	18.5	39.4[b]
AgCl	33.8	26.0	33.8[b]
AgBr	38.2	18.5	38.2[b]
AgI	37.4	9.0	37.4[b]
HF	53.9	59.0	53.9
HCl	24.7	18.5	24.7
HBr	20.8	12.0	20.8
HI	14.1	4.0	14.1
H_2O	46.7	39.0	34.2
H_2S	9.7	4.0	6.1
H_2Se	$-1.2 \approx 0$[c]	2.5	0.0
NH_3	32.2	18.5	16.6
PH_3	7.8	0.0	3.5
AsH_3	$-5.5 \approx 0$[c]	0.5	0.0
BrF	30.1	30.0	30.1
ClF	23.9	22.0	23.9
ICl	8.9	6.5	8.9
IBr	4.0	2.5	4.0

Table 12 (continued)

	Q_{mol}		I_P	Q_a	
NF_3	42.0		22.0	23.6	
NCl_3	$-2.4 \approx 0$ c)		0.0	0.0	
PF_3	63.9		59.0	42.9	
PCl_3	30.8		18.5	16.0	
PBr_3	24.6		12.0	12.2	
PI_3	5.2		4.0	2.3	
AsF_3	66.3		63.0	45.7	
$AsCl_3$	36.1		22.0	19.4	
$AsBr_3$	29.1		15.0	14.8	
AsI_3	16.4		6.5	7.7	
SbF_3	67.3		66.5	46.7	
$SbCl_3$	41.6		26.0	23.3	
$SbBr_3$	38.2		18.5	20.9	
SbI_3	27.8		9.0	14.0	
BiF_3	71.1	(79.0)	66.5	50.4	(61.7)
$BiCl_3$	48.3	(57.4)	26.0	28.0	(36.3)
$BiBr_3$	44.1	(43.2)	18.5	24.6	(24.4)
CF_4	51.7		43.0	25.9	
$CCl4$	9.7		6.5	3.4	
CBr_4	5.6		2.5	1.9	
SiF_4	69.0		70.0	41.8	
$SiCl_4$	39.8		30.0	17.7	
$SiBr_4$	34.0		22.0	14.4	
SiI_4	22.3		12.0	8.5	
$GeCl_4$	37.8		30.0	16.5	
$GeBr_4$	31.5		22.0	13.0	
GeI_4	20.8		12.0	7.9	
$SnCl_4$	41.3		30.0	18.6	
$SnBr_4$	38.2		22.0	16.7	
BF_3	65.3	(62.2)	63.0	43.8	(40.4)
BCl_3	37.6	(31.9)	22.0	19.9	(16.2)
BBr_3	33.3	(26.4)	15.0	17.1	(12.9)
AlF_3	73.4	(90.8)	79.0	53.2	(80.3)
$AlCl_3$	53.6	(57.0)	43.0	32.3	(35.4)
$AlBr_3$	51.8	(48.1)	34.5	30.9	(27.6)
AlI_3	45.9	(36.7)	22.0	25.9	(19.3)
$GaCl_3$		(52.1)	39.0		(31.0)
$GaBr_3$		(42.6)	30.0		(23.6)
GaI_3		(30.3)	18.5		(15.2) d)

For diatomic molecules Q_{mol} is identical with Q_a; values in parentheses are calculated according to the second Pauling postulate.

a) D_{Na-F} and hence Q_a is probably too low.
b) Values Q_a somewhat incertain due to inaccurate thermochemical data.
c) Negative values meaningless, values therefore assumed to be zero.
d) Value too low, as stated in section 4.

a) Diatomic Molecules

The ionic character of the alkali metal halides decreases from fluoride to iodide for a given metal ion, chlorides and bromides having about the same Q_a-values. Q_a values are much less differentiated than the corresponding *Pauling* values. As a consequence, Q_a values for bromides and especially iodides are considerably higher and those for the fluorides lower than the corresponding I_P values.

For a given halogen, both Q_a and I_P values increase with increasing atomic number of the metal, with the exception of NaF, possibly due to a somewhat incorrect D_{Na-F} value.

I_P values for LiI, LiBr, NaI, KI, RbI and CsI are lower than for HF. In contrast, Q_a values for all alkali metal halides are higher than for HF, and this appears more realistic with respect to the physical and chemical properties of the said compounds. The relatively high ionic character of HF is reflected in its strong tendency to association via hydrogen bonds. The ionic character of HCl, HBr and HI is much smaller than that of HF with Q_a values generally higher than I_P.

For the diatomic interhalogen compounds listed in Table 12, I_P and Q_a values are not much different. IBr and ICl are characterized as typical covalent molecules while for ClF and BrF an ionic character $Q_a \approx 24\%$ and 30%, respectively, is calculated. The Trouton constant of 28 e.u. for ClF is indicative of some tendency to association in the liquid state.

The ionic character calculated according to *Pauling* for copper (I)- and silver (I)-halides is surprisingly low. AgBr and CuBr have I_P values similar to that of hydrogen chloride and for AgI I_P amounts to only 9%. In contrast, an ionic character Q_a between 37 and 40% is obtained, and this appears more reasonable.

b) Polyatomic Molecules

For all halogen compounds, both Q_a and I_P decrease from fluoride to iodide. For group IV halides, Q_a values follow the series C ≪ Si > Ge < Sn corresponding to the variation of the Allred-Rochow electronegativities (5)[3]. For halides of group V elements, the order for both I_P and Q_a is

$$N < P < As < Sb < Bi$$

which again follows the Allred-Rochow electronegativities with the exception of As.

---

[3] For further information about the concept of electronegativities see also (6).

Values I_P and Q_a for group III halides all show the same sequence

$$B < Al > Ga$$

again in agreement with the Allred-Rochow scale.

With the exception of NF_3, all fluorides have surprisingly high I_P values; e. g. for SiF_4 the I_P value is similar to that of KCl and RbCl, similarly AsF_3, SbF_3, BiF_3 and BF_3 have about the same I_P values as alkali bromides. Q_a values are considerably lower and this appears more reasonable; for example, SiF_4 ($I_P = 70\%$, $Q_a = 41{,}8\%$) is a gas at room temperature which shows no tendency to association.

Q_a-values for chlorides, bromides and iodides of group IV elements are lower than the Pauling values and are characteristic of predominantly covalent compounds; all these compounds crystallize in molecular lattices and are soluble in organic solvents.

Good agreement is observed between Q_a and I_P for most of the group V chlorides, bromides and iodides, as well as for BCl_3 and BBr_3. I_P values are considerably lower than Q_a for SbI_3 and $BiBr_3$.

BCl_3, BBr_3 and the phosphorous halides are characterized by Q_a as predominantly covalent compounds. The increasing ionic character for arsenic — and especially antimony — and bismuth halides is reflected in the increasing tendency for ionization in the liquid state.

Chlorides, bromides and iodides of aluminium and gallium show considerably higher Q_a values than the boron halides, and this is in agreement with the high tendency of these compounds to form dimers, even in the gas phase[4]. In the solid state, $GaCl_3$, $GaBr_3$, GaI_3, $AlBr_3$ and AlI_3 form molecular lattices composed of dimeric units M_2X_6, whereas AlF_3 in agreement with its considerably higher ionic character crystallizes in a predominantly ionic lattice with octahedrally coordinated aluminium. Q_a values for $AlCl_3$, $GaCl_3$ and $GaBr_3$ are considerably lower than the *Pauling* values.

Reasonable agreement is found between Q_a and I_P values for H_2S, H_2Se, NH_3, PH_3 and AsH_3; all these compounds are characterized as predominantly covalent compounds in agreement with their properties. The higher ionic character of water ($Q_a = 34.2\%$) is reflected in its high tendency to form hydrogen bridges.

Q_a and I_P values are generally too high for compounds in which back-donation occurs, since the bond energy D_{A-B} calculated from thermochemical data will be higher than that of a single bond. Thus, the ionic character of the B—F bond in BF_3 presumably still lies well below

---

4) Note that values of ionic character are valid for the gaseous monomeric planar molecules.

$Q_a = 43.8\%$ (compare with $I_P = 63\%$) and this explains the covalent properties of this compound. Back-donation possibly also occurs in molecules such as PF_3 and SiF_4.

7. Conclusion

Q_a values for the ionic character of bonds calculated from quantities Δ' and D_{A-B} for a large number of diatomic and binary polyatomic molecules appear reasonable with respect to the physical and chemical properties of the compounds. In many cases Q_a values appear to be more reasonable than I_P values calculated according to Pauling from the electronegativities of the bonded atoms. When speaking of the ionic character of a bond, one should always refer to the molecule to which it belongs; thus the ionic character of the $Si-F$ bond will be different in $(CH_3)_3SiF$ and in SiF_4. Reliable Q_a values can be determined only if the homoatomic bond energies of the bonded atoms are known. *Pauling's* assumption, that Δ values of bonds can be determined from the standard enthalpies of formation of compounds, which is used for calculation of electronegativities and can be used for calculation of unknown homoatomic bond energies, is valid only to a limited extent. For instance, applying it to the alkali metal halides yields values which are definitely in error. The same is expected for all compounds which are predominantly ionic in character and show high heats of sublimation, such as the alkaline-earth halides and many metal fluorides (AlF_3, SnF_4 etc.) *Pauling's* assumption can, however, be applied to compounds of nonmetallic elements which have essentially molecular structures in the standard state; for compounds of metallic elements, best values of homoatomic bond energies (and electronegativities) can probably be calculated from enthalpies of formation for the chlorides and bromides.

8. References

1. *Pauling, L.:* The Nature of the Chemical Bond, 3rd ed. Ithaca—New York: Cornell University Press 1960.
2. Gmelins Handbuch der anorganischen Chemie, Chlor, Ergänzungsband B/1. Weinheim/Bergstraße: Verlag Chemie 1968; several references.
3. *Gunn, S. R., Green, L. G.:* J. Phys. Chem. *65*, 178 (1961).
4. *Allred, A. L.:* J. Inorg. Nucl. Chem. *17*, 215 (1961).
   — *Rochow, E. G.:* J. Inorg. Nucl. Chem. *5*, 264 (1958).
6. *Pritchard, H. O., Skinner, H. A.:* Chem. Rev. *55*, 745 (1955).

Received June 7, 1971

Vibrational Spectra and Structural Properties of Complex Tetracyanides of Platinum, Palladium and Nickel

Dr. S. Jerome-Lerutte

Laboratoire de Cristallographie, Université de Liège, Sart Tilman par Liège, Belgium

Table of Contents

1. Introduction

This review analyses the vibrational spectra of complex cyanides and relates them to recently obtained structural data. In no structure is there a strong bond, and there may be no bond at all between the CN ligands and the metal cations. The H bonds are often very weak and when they are stronger, as in organic salts, they do not seem to influence the spectra. This invalidates *Kharitonov's* hypothesis which attempts to ex-

153

plain vibrational spectra, and to an even larger extent electronic spectra, by such interactions with the surroundings of the complex unit.

The proposed force constant calculations are also reviewed. Attention is drawn to some common errors. An approximate calculation is made according to a simple valence force field (SVFF) but experimental data are still too sparse to allow a more precise approach.

The vibrational spectra of complex ions are closely related to their symmetry. In a rough approximation, the group symmetry of the $M(CN)_4^{--}$ ion is D_{4h}, the complex group being square-planar and its center being occupied by the metal atom.

Representations induced by Cartesian displacements of the atoms and selection rules show that six of the sixteen normal vibrations are IR-active and seven are Raman-active.

$$\Gamma M(CN)_4^{--} = 2\,A_{1g} + A_{2g} + 2\,B_{1g} + 2\,B_{2g} + E_g + 2\,A_{2u} + 2\,B_{2u} + 4\,E_u$$

$$\Gamma\mu = A_{2u} + E_u$$

$$\Gamma\alpha = 2\,A_{1g} + B_{1g} + B_{2g} + E_g$$

We shall not describe the internal coordinate system nor the symmetry coordinates, the symmetry force constants and the kinetic energy matrix, as they are detailed in *McCullough's* paper (*1*). *Wilson's* book, "Molecular vibrations", gives further explanations of the methods used (*2*).

The first work on the vibrational spectra of complex cyanides was that of *Mathieu* and *Cornevin* in 1939 (*3*); the Raman spectra of $Na_2M(CN)_4 \cdot 3\,H_2O$ (M = Pt, Pd, Ni) are described and the observed bands are correctly assigned. *Sweeny* and his collaborators (*4*) give the IR spectrum of $K_2\,Pt(CN)_4 \cdot 3\,H_2O$ and propose a preliminary calculation of force constants which we think is wrong; *Bonino* (*5*) and *Salvetti* (*6*) refer to that paper for their assignments of the $Pd(CN)_4^{--}$ spectrum. The calculation of force constants in the $Ni(CN)_4^{--}$ ion in its Na^+ and Ba^{2+} salts was made by *McCullough* and completed by *Jones* (*7*) in the frequency range below 300 cm^{-1}. A simple valence force field (SVFF) is used. Our calculations refer to both these studies.

Kharitonov and collaborators (*8—12*) describe IR spectra of many tetracyanoplatinates, whether water-soluble or not. They explain band shifts and splittings by the presence of strong bonds between ligands and cations or their hydration sphere. We shall discuss this work later, pointing out the structural pecularities of each compound.

This short summary would not be complete if it did not mention *L. H. Jones's* work on complex cyanides; linear as $Au(CN)_2^-$ (*13*), $Ag(CN)_2^-$ (*14—15*), quadratic as $Au(CN)_4^-$ (*16*) or octahedral as $Co(CN)_6^{3-}$ and $Fe(CN)_6^{3-}$ (*17—20*). He has calculated force constants for $Au(CN)_4^-$ in a simple valence force field.

2. Recording of the Spectra

We have determined the crystal structures of $(C_2H_5NH_3)_2\{_{Pd}^{Pt}\}(CN)_4$ (abbreviated to Et_IPt or Et_IPd), $[(C_2H_5)_2NH_2]_2\{_{Pd}^{Pt}\}(CN)_4$, ($Et_{II}Pt$ or $Et_{II}Pd$), $[(C_2H_5)_3NH]_2\{_{Pd}^{Pt}\}(CN)_4 \cdot 2\ H_2O$ ($Et_{III}Pt$ or $Et_{III}Pd$) by X-Ray methods, and at the same time we have recorded their IR spectra and attempted to describe them in relation to the crystal structure data. Outside the fundamental ranges of the complex ion, *i.e.* between 2000 and 2200 cm^{-1} for ν_{CN} stretching frequencies, and below 600 cm^{-1} for ν_{MC} stretching frequencies and deformation frequencies, intense bands of organic cations obscure the possible combinations and overtones. Thus we have limited our investigations to these parts of the spectra; they have been recorded on a Pe 621 spectrophotometer using both nujol mull and KBr disk methods.

3. Selection Rules and Space Group Symmetry

In a crystalline lattice, the true complex group symmetry is different from but often near to the theoretical D_{4h} one. It can reach at minimum the symmetry of the site occupied by the metal atom in each space group. Selection rules are modified in consequence, but generally their influence on the spectra is limited to the appearance of forbidden bands and the splitting of degenerate ones. The analysis of those spectral features is valuable as it provides information in each particular case on the structure of the complex group.

4. Assignment of ν_{CN} Stretching Frequencies

According to selection rules for the molecular groups, two ν_{CN} vibrations are Raman-active (A_{1g}, B_{1g}) and one is IR-active (E_u). Site symmetry leads to a new distribution of the bands. If the complex group occupies a non-centered site, A_{1g} and B_{1g} frequently appear in the IR spectra. Likewise, the splitting of the E_u line is seen in most of them: It looks like a shoulder on the intense ν_{CN} bond or a complete separation into two adjacent bands.

Disregarding any external influence, the $\nu_{CN}(E_u)$ frequency depends on the ratio n between coordination number and formal charge on the central atom (27). In $Pt(CN)_4^{--}$, $Pd(CN)_4^{--}$ or $Ni(CN)_4^{--}$, this ratio is equal to 2 and ν_{CN} is located at 2130 cm^{-1}, as *Kharitonov's* spectrum of $K_2Pt(CN)_4$ in diluted aqueous solution shows (*8*). According to the same author (*9—12*), there is a shift towards higher frequencies when the complex group is in the crystalline state. He explains this by the formation of bridging bonds between complex groups and their surroundings, with a decrease in the charge hold by the ligands and a consequent decrease in the coulombic repulsion between square planes. Thus, strong bonds with cations or their hydration spheres would favour the nearness of the heavy atoms with its well-known influence on the electronic spectra.

Let us see how far this theory is verified by known structural studies.

a) Tetracyanoplatinic(II) and Tetracyanopalladic(II) Acids

Although their crystalline structures are not known, some authors assume that strong symmetrical N... H... N bonds are formed, as suggested by the ν_{NH} shift towards lower frequency (*22*). The ν_{CN} stretching band is displaced to 2202 cm^{-1} and the Pt complex is brightly coloured, which seems to agree with *Kharitonov's* hypothesis.

b) Alkaline Salts

Three structures of hydrated compounds are known: $Na_2Pt(CN)_4 \cdot 3\,H_2O$ and its isotypes $Na_2Pd(CN)_4 \cdot 3\,H_2O$ and $Na_2Ni(CN)_4 \cdot 3\,H_2O$ (*23*); $Rb_2Pt(CN)_4 \cdot H_2O$ and its isotype $Rb_2Pd(CN)_4 \cdot H_2O$ (*24—25*) and $Rb_2Pt(CN)_4 \cdot 1.5\,H_2O$ (*26*). The only structure of an anhydrous complex is that of $K_2Ni(CN)_4$ (*27*); $K_2Pt(CN)_4$ is probably isotypic.

Average values as well as lower and upper limits of M'... N distances (M' being alkali metal, N ligand nitrogen atom) in each structure, are given in Table 1. There is a linear relation between the observed average distances and the ionic radius of the cation (Fig. 1).

Table 1 shows that there is no strong bond of the C≡N—M' type in most of the structures studied as N...M' distances are distributed in a narrow range about the mean value which is nearly the sum of the Van der Waals radius of the N atom and the ionic radius of the metal atom. The only exception is for $Rb_2Pt(CN)_4 \cdot 1.5H_2O$ where the minimum Rb—N distance is 0.66 Å shorter than the average one, but *Dupont* (*26*) makes reservations as to the accuracy of the light atom positions in presence of heavy metals like Pt and Rb$^+$. Moreover, this shortening is not related to a shift of the $\nu_{CN}(E_u)$ band. On the other hand, the comparison of the two forms of Rb$^+$ salt, shows that the possible shortening of the N...M' distance could be parallel to that of the M...M distance in the column of complex ions. We shall see later if a relationship between the two facts is suggested by other structural examples.

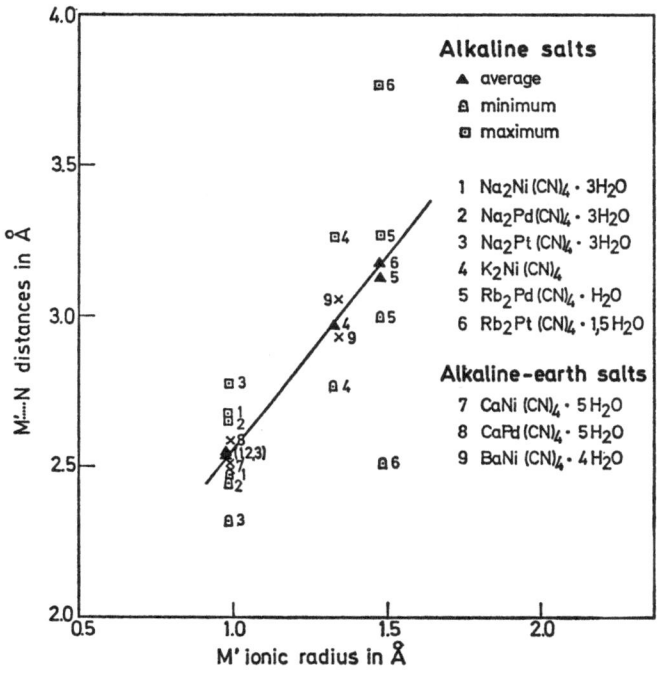

Fig. 1

Table 1

Complex	Na$_2$Ni(CN)$_4$ · 3 H$_2$O	Na$_2$Pd(CN)$_4$ · 3 H$_3$O	Na$_2$Pt(CN)$_4$ [a]) · 3 H$_2$O	K$_2$Ni(CN)$_4$	Rb$_2$Pd(CN)$_4$ · H$_2$O	Rb$_2$Pt(CN)$_4$ · 1.5 H$_2$O[a])
M′...N$_{av.}$	2.54	2.53	2.52	2.97	3.13	3.17
M′...N$_{min.}$	2.44	2.46	2.31	2.76	2.99	2.51
M′...N$_{max.}$	2.67	2.65	2.78	3.26	3.26	3.78
M′ionic radius	0.97	0.97	0.97	1.33	1.47	1.47
M...M distance	3.77	3.73	3.70	4.29	3.72	3.39

			id.	K$_2$Pt(CN)$_4$	Rb$_2$Pt(CN)$_4$ · H$_2$O	id.
Colour			Colourless	White	Colourless	green
$\nu_{CN}(E_u)$			2133—36	2130	2126—2135	2126—2135

[a]) As the authors point out, large deviations are not significant as the presence of heavy Pt atoms considerably diminishes the accuracy of X-Ray structural methods.

As to H bonds in alkaline tetracyanoplatinates, we may say that they are weak:

$$O-H\ldots N_{min}\ (Na_2Pt(CN)_4 \cdot 3\,H_2O) = 2.97\ \text{Å}$$

$$O-H\ldots N_{min}\ (Rb_2Pt(CN)_4 \cdot H_2O) = 3.02\ \text{Å}$$

except for $Rb_2Pt(CN)_4 \cdot 1.5\,H_2O$ which contains two strong bonds: 2.45 and 2.61 Å. The IR spectrum shows a very weak band between 2500 and 2600 cm^{-1} which may be assigned to a shifted OH frequency, but no modification appears in the ν_{CN} range and doubt remains concerning the accuracy of light atom positions.

In the particular case of alkaline salts, the structural studies reveal the weakness of *Kharitonov's* hypothesis. According to him, it is necessary to distinguish between the influence of the decrease in local symmetry, which leads to the appearance of at most four bands, and the distortion of the complex group by bridging bonds with the outer sphere cations or water molecules, which creates new intense bands at higher frequencies Such a distinction regarding the nature of the effect has no physical meaning, as it is evident that a lowering of the site symmetry is always related to the influence of the surroundings. At most, the magnitude of the perturbation and its consequences may be discussed.

If the formation of strong bonds significantly disturbs the symmetry and if D_{4h} approximation is no longer possible, a new theoretical approach to the problem becomes necessary. Contrary to *Kharitonov's* opinion (9), we do not think that the spectra of hydrated alkaline salts are sufficiently disturbed to demand such a calculation. There is no difficulty in assigning frequencies in the $Na_2Pt(CN)_4 \cdot 3\,H_2O$ spectrum on the assumption of a site modification of the D_{4h} symmetry, as has been done by *Mc Cullough* for the Ni complex (Table 3).

c) Organic Salts

The ν_{CN} range is not very crowded in the IR and Raman spectra of the ethylsubstituted ammonium salts whose structures have been studied (Et$_I$Pt, Et$_{II}$Pt, Et$_{III}$Pt) (28—29). The allowed bands are sharp; some splittings and the appearance of weak bands are easily explained by the lowering of local symmetry. Thus, the A_{1g} band appears in the IR spectrum of Et$_I$Pt because the Pt atom is not on a symmetry center. In the spectra of Et$_{II}$Pt and Et$_{III}$Pt, $\nu_{CN}(E_u)$ is slightly shifted towards higher frequencies: 2142 and 2143 cm^{-1} against 2137 cm^{-1} for Et$_I$Pt, and 2130 cm^{-1} for Pt(CN)$_4^{--}$ in solution. Apart from this shift, the shape of the two spectra is very similar to that of the solution and is simpler than for alkaline salts. The known structures show that the quadratic groups are very near to the D_{4h} theoretical symmetry.

As to whether the slight positive shift is due to the presence of fairly strong H bonds with the ammonium ions, the magnitude of the shift incites us to consider its causes with great care.

$Et_{II}Pd(Pt)$ $N-H...N = 2.82 \pm 0.02$ Å
(average value = 3,10 $\pm$ 0.13) (30)

$Et_{III}Pd(Pt)$ $O-H...N = 2.76 \pm 0.05$ Å
(average value 2.88 $\pm$ 0.13 Å) (30)

d) Alkaline Earth Salts

Some of these salts, namely Mg $Pt(CN)_4 \cdot 2$ H_2O, $MgPt(CN)_4 \cdot 7$ H_2O and anhydrous salts in general, show spectra which are much modified from that of the isolated ion. They have strong bands, well-defined or overlapping, in the ν_{CN} frequency range, and the intensity of the higher frequency component is difficult to explain by a mere lowering of local symmetry. Unfortunately, none of these structures is known and no conclusion can be drawn.

The structures which have been determined are those of $CaNi(CN)_4 \cdot$ 5 H_2O (31), $CaPd(CN)_4 \cdot 5$ H_2O (32) and $BaNi(CN)_4 \cdot 4$ H_2O (33). The isotypism between Ni, Pd and Pt compounds allows us to make an extrapolation. Structural data are collected in Table 2.

The cation to ligand distances in Ca^{2+} and Ba^{2+} salts depend on the ionic radii and exhibit the same relationship as in the alkaline salts (Fig. 1). Thus, there is no possibility of strong cation-ligand bonds in these alkaline-earth salts.

Table 2

Complex	$CaNi(CN)_4$ $\cdot 5$ H_2O	$CaPd(CN)_4$ $\cdot 5$ H_2O	$BaNi(CN)_4$ $\cdot 4$ H_2O
M'...N (1)	2.487 (5)	2.58	2.93 (1)
M'...N (2)	2.509 (5)	2.59	3.05 (1)
M' ionic radius	0.99	0.99	1.34
M...M	3.39	3.42	3.36
		$CaPt(CN)_4$ $\cdot 5$ H_2O	$BaPt(CN)_4$ $\cdot 4$ H_2O
Colour		yellow	yellow
$\nu_{CN}(E_u)$		2124—56 cm^{-1}	2127—36 cm^{-1}

159

As to the observed H bonds ,they are weaker than those measured in $Et_{III}Pt$:

$$OH \ldots N_{min} \, (CaNi(CN)_4, \, 5 \, H_2O) = 2.93 \, \text{Å}$$

$$OH \ldots N_{min} \, (CaPd(CN)_4, \, 5 \, H_2O) = 2.94 \, \text{Å} \, .$$

The splitting of the $\nu_{CN}(E_u)$ band is $\Delta\nu = 9$ cm^{-1} in the Ba^{2+} salt and $\Delta\nu = 32$ cm^{-1} in the salt of Ca^{2+}. This can be explained by symmetry effects but it is surprisingly larger in CaPt(CN)$_4 \cdot 5$ H$_2$O than in Na$_2$Pt (CN)$_4 \cdot 3$ H$_2$O although distortion from the theoretical D$_{4h}$ symmetry is greater in the latter case.

e) Transition Metal Salts

All spectra show intense bands beyond 2170 cm^{-1} (32). *Kharitonov* assumes polymeric structures with strong N → M' bonds. Such compounds are marginal for this approach, as their crystalline structures and their electronic spectra are inaccessible by present techniques.

f) Discussion

In most cases, the presence of a high-frequency component among the IR ν_{CN} bands is not related to the formation of strong bonds between ligands and cations, nor between ligands ans water molecules. *Kharitonov's* assumptions, which are weakened by structural studies, are based upon *Dow's* work (27), but this deals with the ν_{CN} band position in spectra of compounds quite different in their symmetry. A distinction must be made between D$_{4h}$ and O$_h$ complex groups and to an even larger extent between those and polymeric structures with M—CN—M' bridges, like Pd(CN)$_2$ or Ni(CN)$_2$. Each particular case exhibits a new force field needing a special calculation.

The situation is quite different here, the complex group keeping its approximate D$_{4h}$ symmetry in every case. The influence of the surroundings is limited to minor shifts and splittings.

In some alkaline-earth and transition-metal salts, the structures of which are unfortunately unknown, differences appear which can hardly be explained by mere distortions but which may indicate deeper structural changes.

Nevertheless, the stacking of complex ions in columns has no effect on the spectra, as there appears to be no close relationship between the band position and the distance between heavy metal atoms.

5. M—C Vibrations

Apart from the ν_{CN} range, frequency assignments in complex tetra-cyanides are sometimes unreliable. *Jones's* assignments for $Au(CN)_4^-$ and *Mc Cullough's* assignments for $Ni(CN)_4^{--}$ are correct. Indeed, they describe all the observed bands, and the calculated force constants are consistent with those in other complexes of similar type, such as linear or octahedral cyanides, thiocyanates and carbonyls. The frequencies taken from these two papers are given in Table 3.

The assignments and subsequent calculations of $Pt(CN)_4^{--}$ (4) and of $Pd(CN)_4^{--}$ (6) are not consistent with the previous ones. In particular, the deformation force constants are abnormally weak. Thus, we propose a new assignment in analogy to $Ni(CN)_4^{--}$ and $Au(CN)_4^-$, but some caution must be observed in a frequency range where internal deformations and external lattice vibrations can be confused.

a) Pt—C Vibrations

Raman frequencies (A_{1g}, B_{1g}) are doubtless 465—455 cm^{-1} in $Na_2Pt(CN)_4$ · 3 H_2O and 486—461 cm^{-1} in Et_IPt. Our study has been limited to Et_IPt for technical reasons, but it is certain that comparison of the ν_{MC} frequency in different salts would yield important information. The distinction between the two normal modes is based upon the polarization factor and the low-frequency approximation:

$$\lambda(A_{1g}) / \lambda(B_{1g}) > 1.$$

The IR-active E_u representation is assigned to 500 cm^{-1}. The band is sharp in alkaline and organic salts but it appears more blurred in most of the alkaline-earth and transition-metal salts. It seems to shift towards a lower frequency when ν_{CN} shifts towards higher ones, but the available information is not sufficient to enable us to draw any conclusions.

b) Pd—C Vibrations

$\nu(A_{1g})$ and $\nu(B_{1g})$ almost coincide at 439 cm^{-1} in the Na^+ salt and are well separated (423 and 447 cm^{-1}) in Et_IPd. As $G(A_{1g})$ and $G(B_{1g})$ matrices are independent of the heavy atom, the decrease of the F matrix:

$$F_{Pt\,C} > F_{Pd\,C} > F_{Ni\,C}$$

involves the same order of frequencies. Unfortunately the $G(E_u)$ matrix varies in inverse ratio to the $F(E_u)$ matrix and prevents the evaluation of the IR frequency range. *Bonino* observes it at 495 cm^{-1} in Na^+ and K^+ salts; nothing contradicts this assignment.

Table 3

	Ni(CN)$_4^{--}$		Pd(CN)$_4^{--}$		Pt(CN)$_4^{--}$		Au(CN)$_4^{--}$	Activity
	McCullough (1960) Na$_2^+$	D. Jones (1968) Na$_2^+$	Bonino (1959) K$_2^+$	Author (1970) Et$_4^+$	Sweeny (1956) K$_2^+$	Author (1970) Et$_4^+$	L. H. Jones (1964) K$_2^+$	
$\nu_{CN}(A_{1g})$	2149	2149	2160	2175	2168	2178	2207	R
$\nu_{MC}(A_{1g})$	419[a]	417	437	447	465	486	459	R
$\nu_{MCN}(A_{2g})$	325[a]	468	—	—	—	—	—	R
$\nu_{CN}(B_{1g})$	2141	2141	2145	2158	2149	2155	2198	R
$\nu_{MC}(B_{1g})$	405[a]	401	—	423	455	461	450	R
$\nu_{MCN}(B_{2g})$	488[a]	490	290	399	318	427	422	R
$\nu_{CMC}(B_{2g})$	(91)	—	92	—	95	—	110	R
$\nu_{CN}(E_u)$	2132 / 2128	2132 / 2128	2140	2138	2150	2137 / 2120	2189	I–R
$\nu_{MC}(E_u)$	543	543	495	—	505	500	462	I–R
$\nu_{MCN}(E_u)$	433 / 421	433 / 421	280[a]	388 / 380	300	427 / 413	415	I–R
$\nu_{CMC}(E_u)$	(78)	—	80[a]	—	—	—	—	I–R
$\nu_{MCN}(A_{2u})$	448	448	415	410 ?	410	470	—	I–R
$\nu_{MCN}(A_{2u})$	(77)	—	—	—	—	—	—	I–R
$\nu_{MCN}(B_{2u})$	(303)	—	—	—	—	—	—	I–R
$\nu_{CMC}(B_{2u})$	(54)	—	—	—	—	—	—	I–R
$\nu_{MCN}(E_g)$	280	301 / 307	390	297	—	324	298	R

[a] From combination or overtone bands.

6. MCN Deformations

In-plane deformations are distributed among B_{2g}(R) and E_u(IR) representations and out-of-plane ones have A_{2u}(IR) and E_g(R) symmetry.

a) Pt—CN Deformations

All IR spectra of alkaline and organic $Pt(CN)_4^{--}$ salts have an intense band in the 420 cm^{-1} range which is often split. No doubt, it is δ_{PtCN} (E_u) which is found to be intense at almost the same place in other quadratic cyanides. *Sweeny's* assignment at 300 cm^{-1} is wrong, as no band appears at this frequency and the calculated deformation constant is too weak.

δ_{PtCN} (B_{2g}) is observed in the Raman spectrum at 430 cm^{-1} and not at 318 cm^{-1}, supporting the approximation based upon the order of F and G matrices

$$\delta_{PtCN} (B_{2g}) \; > \; \delta_{PtCN} (E_u)$$

This band (312 cm^{-1}) is the out-of-plane deformation $\pi_{PtCN}(E_g)$. The approximation π_{PtCN} (A_{2u}) $> \pi_{PtCN}$ (E_g) may be justified by a weak band at 470 cm^{-1} but its assignment remains uncertain.

b) Pd—CN Vibrations

We do not agree with *Salvetti's* assignment. On the analogy of the Et$_I$Pt spectrum, δ_{PdCN} (E_u), strong and split, is at 390 cm^{-1}; δ_{PdCN} (B_{2g}) is at 399 cm^{-1} and π_{PtCN} (E_g) at 297 cm^{-1}. A very weak band, which might be the $\pi(A_{2u})$ deformation, appears at 410 cm^{-1} in the IR spectra of Et$_I$Pd and Et$_{II}$ Pd.

7. CMC Deformations

These frequencies are low in the spectrum and inaccessible to our spectrophotometer. Raman spectra cover this range but their resolution is poor close to the exciting line. Thus, our approach takes provisional account of the frequencies observed by *Mathieu*: δ_{CPtC} (B_{2g}) $= 95$ cm^{-1} and δ_{CPdC} (B_{2g}) $= 94$ cm^{-1}.

Lorenzelli and *Delorme* (34) have observed bands in the far IR spectra at 180 cm^{-1} and 140 cm^{-1}, but they cannot explain them. The hypothesis of deformation fundamentals is unacceptable as it leads to inconsistencies in the force constant calculation.

8. Secular Equation

The secular equation of vibrating motion can be described by a general equation:

$$GF - E\lambda = 0$$

where E is a unitary matrix.

The most general potential function leads to a 16-dimensional matrix which contains 27 independent force constants and makes approximations absolutely necessary.

The approximation of the diagonal valence force field (DVFF) function arbitrarily neglects off-diagonal terms and estimates just 16 diagonal force constants but, as *McCullough* notes, it is a mathematical approach without any physical meaning.

The SVFF approach is rougher but physically more significant; it neglects all interactions and gives six force constants: F_r (CN), F_R (MC) F_α (MCN in-plane), F_β (CMC in-plane) F_φ (MCN out-of-plane), F_ϑ (CMC out-of-plane). In the $Ni(CN)_4^{--}$ case, the two potential functions have led to almost similar values. The Urey-Bradley potential function is more sophisticated and has sometimes been used, but it only takes into account the repulsion between neighbouring C or N atoms, which is not the main one. All results from the literature and from our own work are compared in Table 4.

Table 4

	$Ni(CN)_4^{--}$	$Pd(CN)_4^{--}$		$Pt(CN)_4^{--}$		$Au(CN)_4^-$
	SVFF	UB	SVFF	UB	SVFF	SVFF
F_r	16.67	16.70	16.73	16.82	16.70	17.44
F_R	2.60	2.90	2.90[a]	3.42	3.41	2.99
F_α	0.42	0.07	0.35	0.08	0.43	0.42
F_β	(0.23)?	0.06	0.32	0.08	0.32	0.51
F_φ	0.34	0.17	0.25		0.30	0.30
F_ϑ	0.12	0.10	[b]		[b]	0.51

[a]) Assumed, as $\nu_{MC}(E_u)$ is unknown.
[b]) Unknown, as $\nu_{CMC}(A_{2u})$ is not assigned; in calculation, F_φ is assumed equal to F_ϑ.

F_r and F_R are given in mdyne $Å^{-1}$ and the other constants in mdyne $Å$ rad^{-2}.

9. Force Constants

As expected by Dow's rule, F_r (CN) has approximately the same value in Pt(II), Pd(II), Ni(II) complexes, whereas F_R constants vary greatly from one symmetry representation to another; this assumes important interactions between metal-carbon vibrations.

Deformation constants are withdrawn to reasonable limits and the agreement with similar complexes allows previous assignments for Pt(CN)$_4^{--}$ and Pd(CN)$_4^{--}$ to be rejected. Nevertheless, we hope to study the question further by the application of a more specific potential function and to localize definitely the unknown normal modes in the range below 500 cm^{-1}.

10. Conclusions

Our aim is to record the state of the art in the vibrational analysis of tetracyanide complex spectra.

In the first part, we discuss *Kharitonov's* theory which finds a parallelism between the evolution of the ν_{CN} stretching bands and the formation of strong bonds between ligands and their surroundings. We have shown that this theory is not corroborated by structural data. Unfortunately, the difficulty of obtaining accurate structural measurements of distances and the scarcity of such results makes our demonstration less clear than we should have desired. Nevertheless, there is no doubt that, even if such bonds are sometimes possible, they have no direct influence on the vibrational spectra. Most of the spectral features are quite satisfactorily explained by decreased site symmetry.

In the second part, we check the assignments of frequencies in the spectra of M(CN)$_4$ ions. The proposed assignments are correct for Ni(CN)$_4^{--}$ and Au(CN)$_4^{-}$ but not for Pd(CN)$_4^{--}$ and Pt(CN)$_4^{--}$. In several papers, *Jones* has suggested new assignments for Pt complexes, but in spite of that, some authors (*35*) continue to adopt *Sweeny's* description. This is the reason why we have included the discussion of that point in our review.

Acknowledgments. I am indebted to Professor *H. Brasseur* and to Professor *J. Toussaint* for their helpful assistance. I also thank my colleagues of the crystallography laboratory for giving me access to their structural data prior to publication.

11. References

1. *McCullough, R. L., Jones, L. H., Crosby, G. A.:* Spectrochim. Acta *16*, 929 (1960).
2. *Wilson, E. B., Jr., Decius, J. C., Cross, P. C.:* Molecular Vibrations. New York: McGraw-Hill 1955.
3. *Mathieu, J. P., Cornevin, S.:* J. Chim. Phys. *56*, 271 (1939).
4. *Sweeny, D. M., Nakagawa, I., San-Ichiro Mizushima, Quagliano, I. V.:* J. Am. Chem. Soc. *78*, 889 (1956).
5. *Bonino, G. B., Chiorboli, P., Fabbri, G.:* Rend. Accad. Nazl. Lincei *26*, 137 (1959).
6. *Salvetti, O.:* Rend. Accad. Nazl. Lincei *26*, 225 (1959).
7. *Jones, D., Hyams, I. J., Lippincott, E. R.:* Spectrochim. Acta *24*A, 973 (1968).
8. *Kharitonov, Yu. Ya., Evstaf'eva, O. N., Baranovskii, I. B.:* Russ. J. Inorg. Chem. (English Transl.) *11*, 1354 (1966).
9. — — — *Mazo, G. Ya.:* Russ. J. Inorg. Chem. (English Transl.) *13*, 434 (1968).
10. — Russ. J. Inorg. Chem. (English Transl.) *14*, 403 (1969).
11. — — *Baranovskii, I. B., Mazo, G. Ya.:* Russ. J. Inorg. Chem. (English Transl.) *14*, 530 (1969).
12. — — — — Russ. J. Inorg. Chem. (English Transl.) *14*, 1108 (1969).
13. *Jones, L. H.:* J. Chem. Phys. *27*, 468 (1957).
14. — J. Chem. Phys. *26*, 1578 (1957).
15. — Spectrochim. Acta *19*, 1675 (1963).
16. — *Smith, J. M.:* J. Chem. Phys. *41*, 2507 (1964).
17. — J. Chem. Phys. *36*, 1209 (1962).
18. — J. Mol. Spectry. *8*, 105 (1962).
19. — Inorg. Chem. *2*, 777 (1963).
20. — J. Chem. Phys. *41*, 856 (1964).
21. *Dows, D. A., Haim, A., Wilmarth, W. K.:* J. Inorg. Nucl. Chem. *21*, 33 (1961).
22. *Evans, D. F., Jones, D., Wilkeinson, G.:* J. Chem. Soc. 3164 (1964).
23. *Ledent, J.:* Thesis, Univ. Liège (1970).
24. *Dupont, L.:* Bull. Soc. Roy. Sci. Liège *36*, 47 (1967).
25. — Acta Cryst. B. *26*, 964 (1970).
26. — Bull. Soc. Roy. Sci. Liège *38*, 509 (1969).
27. *Vannerberg, N. G.:* Acta Chem. Scand. *18*, 2385 (1964).
28. *Jerome-Lerutte, S.:* Acta Cryst. (in press) (1971).
29. — Thesis, Univ. Liège (1970).
30. *Pimentel, G. C., McLellan, A. L.:* The Hydrogen Bond. San Francisco: Freeman and Co 1960.
31. *Holt, E. M., Watson, K. J.:* Acta Chem. Scand. *23*, 14 (1969).
32. *Fontaine, F.:* Bull. Soc. Roy. Sci. Liège *9—10*, 437 (1968).
33. *Larsen, K. F., Hazell, R. G., Rasmussen, S. E.:* Acta Chem. Scand. *23*, 61 (1969).
34. *Lorenzelli* and *Delorme:* Spectrochim Acta *19*, 2033 (1963).
35. *Nakamoto, K.:* (1970) IR. Spectra of Inorganic and Coordination Compounds. New York: Wiley-Interscience 1970.

Received June 14, 1971

Electronic Spectra and Structural Properties of Complex Tetracyanides of Platinum, Palladium and Nickel

M. L. Moreau-Colin

Laboratoire de Cristallographie, Université de Liège au Sart Tilman, Liège, Belgium

Table of Contents

I. Introduction

Radiocrystallographic studies of compounds containing the planar complexes $Pt(CN)_4^{--}$, $Pd(CN)_4^{--}$ and $Ni(CN)_4^{--}$ have shown the general[1] isotypism of tetracyanides which derive from the same cation.

---

[1] *F. Fontaine, M. L. Moreau* and *J. Simon* (*1*) have pointed out that the platinocyanides have c axes which are systematically shorter than those of the pallado and nickel cyanides.

K. Krogmann (*2*) confirms the same characteristic and stresses that the platinocyanides are the only compounds which are able to superpose their plane $Pt(CN)_4^{--}$ at a shorter distance than 3.25 Å. He notes that, when this distance is really shorter than 3.25 Å, there is no isotypism of platinocyanides with the two analogue complexes of Pd and Ni.

Table 1

	System	z	$a_{(Å)}$	$b_{(Å)}$	$c_{(Å)}$
$BaNi(CN)_4$, $4\ H_2O$	monoclinic	4	12.07	13.61	6.72
$BaPd(CN)_4$, $4\ H_2O$	monoclinic	4	11.98	13.83	6.73
$BaPt(CN)_4$, $4\ H_2O$	monoclinic	4	11.89	14.08	6.54
$CaNi(CN)_4$, $5\ H_2O$	orthorhombic	8	17.13	18.76	6.77
$CaPd(CN)_4$, $5\ H_2O$	orthorhombic	8	17.33	19.29	6.84
$CaPt(CN)_4$, $5\ H_2O$	orthorhombic	8	17.36	19.30	6.72
$SrNi(CN)_4$, $5\ H_2O$	monoclinic	4	10.35	15.21	7.29
$SrPd(CN)_4$, $5\ H_2O$	monoclinic	4	10.64	15.51	7.26
$SrPt(CN)_4$, $5\ H_2O$	monoclinic	4	10.58	15.45	7.15
$Rb_2Pd(CN)_4$, $1\ H_2O$	orthorhombic	4	10.01	13.74	7.44
$Rb_2Pt(CN)_4$, $1\ H_2O$	orthorhombic	4	9.99	13.61	7.37
$Na_2Ni(CN)_4$, $3\ H_2O$	triclinic	4	15.22	8.93	7.46
$Na_2Pd(CN)_4$, $3\ H_2O$	triclinic	4	15.403	9.049	7.388
$Na_2Pt(CN)_4$, $3\ H_2O$	triclinic	4	15.43	9.05	7.34
$Rb_2Pd(CN)_4$, $1.5\ H_2O$	monoclinic	8	—	—	—
$Rb_2Pt(CN)_4$, $1.5\ H_2O$	monoclinic	8	12.67	12.78	13.57
$[C_2H_5NH_3]_2\ Pd(CN)_4$	tetragonal	16	17.15	—	18.90
$[C_2H_5NH_3]_2\ Pt(CN)_4$	tetragonal	16	17.01	—	18.75
$[(C_2H_5)_2NH_2]_2\ Pd(CN)_4$	triclinic	2	15.79	9.10	6.35
$[(C_2H_5)_2NH_2]_2\ Pt(CN)_4$	triclinic	2	15.85	9.06	6.35
$[(C_2H_5)_3NH]_2\ Pd(CN)_4$, $2\ H_2O$	monoclinic	2	16.54	9.14	8.22
$[(C_2H_5)_3NH]_2\ Pt(CN)_4$, $2\ H_2O$	monoclinic	2	16.55	9.12	8.17

The results shown in Table 1 enable comparisons to be made which confirm this isotypism.

The optical characteristics of these compounds are, however, quite different. The optically negative nickel cyanides are all yellowish-brown and the palladocyanides are optically negative and colourless (Table 1).

The platinocyanides have various colours, are positive or negative according to the nature of the cation, and are characterized by a very intense luminescence which also varies according to the nature of the cation and the degree of hydration.

Crystallographers and physicists have long studied the differences in the behaviour of these isotypic compounds and many attempts have been made to explain them. We should like to show where these differences are real or apparent and to what extent the nature and the involvement of the plane $M(CN)_4^-$ in the crystal modify the absorption and the emission of the same ion in the "free" state.

α_0	β_0	γ_0	Space-groups	optical sign.	Colour	Ref.	Fluorescence
—	107° 54'	—	C 2/C	—	orange	(3)	green
—	104° 28'	—	C 2/C	—	colourless	(4)	
—	103° 54'	—	C 2/C	+	green		
—	—	—	P cab	—	orange	(5)	blue
—	—	—	P cab	—	colourless		
—	—	—	P cab	+	green		
—	99°	—	C 2/N	—	orange	(6)	violet
—	97° 17'	—	C 2/N	—	colourless		
—	95° 7'	—	C 2/N	—	colourless		
—	—	—	P ncn	—	colourless	(7)	violet
—	—	—	P ncn	+	colourless		
95.43°	92.54°	89.13°	p $\bar{1}$	—	orange	(8)	violet
95.28°	92.38°	89.30°	p $\bar{1}$	—	colourless		
95.13°	92.39°	89.07°	p $\bar{1}$	—	colourless		
—	—	—	C 2/C	—	colourless	(9)	blue
—	99° 48'	—	C 2/C	+	green		
—	—	—	I 41/acd	—	colourless	(10)	
—	—	—	I 41/acd	+	colourless		
84° 11'	92° 42'	94° 5'	C $\bar{1}$ ou C 1	—	colourless	(10)	
83° 7'	93° 1'	94°	C $\bar{1}$ ou C 1	—	colourless		
—	93° 32'	—	C2/m, C$_2$ ou Cm —		colourless	(10)	
—	93° 34'	—	C2/m, C$_2$ ou Cm —		colourless		

2. General Views on the Crystalline Structure of the Tetracyanides of Pt, Pd and Ni

Radiocrystallographic studies of these tetracyanides have shown a general property suggested by the negative optical character and the strong birefringence of the crystals examined. It is known that, even in the case of compounds formed by very anisotropic groups, such as the planar group $M(CN)_4^-$, the negative optical character and the strong birefringence can only be explained if, in the structure, the planar groups are all parallel with one another. The study of crystalline structure generally confirms this conclusion, the $M(CN)_4^-$ groups being stacked one above the other in columns in the direction of the shortest axis (which is the c axis) at the rate of 2 groups per length c. It can be seen that this c axis varies from one compound to another, and in 1938. *H. Brasseur* advanced the hypothesis that the $M(CN)_4$ groups situated in a column can

169

undergo a rotation around the axis of the column. The smaller this axis, the larger the angle of the relative rotation.

Table 2 presents the most recent crystallographic results concerning

Table 2

Compounds		$BaNi(CN)_4$, 4 H_2O	$CaNi(CN)_4$, 5 H_2O	$Na_2Pd(CN)_4$, 3 H_2O			
Reference		(3)	(11)	ion n⁰ 1 (8)		ion n⁰ 2	
Distances (Å)	M—C$_1$	1.859 (7)[a]	1.859 (6)	1.978 (11)	1.980[b]) (11)	1.969 (11)	1.975[b]
	M—C$_2$	1.861 (7)	1.858 (6)	2.001 (11)	2.007[b]) (11)	1.975 (11)	1.979[b]
	M—C$_3$		1.860 (6)	2.033 (12)	2.040[b]) (12)	2.017 (12)	2.026[b]
	M—C$_4$		1.863 (6)	1.988 (11)	1.993[b]) (11)	1.981 (13)	1.990[b]
Distances (Å)	C$_1$—N$_1$	1.145 (10)	1.154 (8)	1.182 (15)	1.202[b]) (15)	1.162 (15)	1.171[b]
	C$_2$—N$_2$	1.163 (10)	1.152 (9)	1.139 (15)	1.150[b]) (15)	1.171 (14)	1.187[b]
	C$_3$—N$_3$	1.163 (10)	1.152 (16)	1.129 (16)	1.155[b]) (16)	1.132 (17)	1.146[b]
	C$_4$—N$_4$	1.163 (10)	1.133(8)	1.130 (15)	1.144[b]) (15)	1.145 (17)	1.151[b]
Angles (degree)	M—C$_1$—N$_1$	177.3 (8)	179.4 (6)	176.8 (1.0)		175.4 (1.1)	
	M—C$_2$—N$_2$	178.8 (7)	177.3 (6)	174.9 (1.1)		174.2 (1.1)	
	M—C$_3$—N$_3$	178.8 (7)	176.4 (5)	178.8 (1.3)		174.7 (1.4)	
	M—C$_4$—N$_4$	178.8 (7)	177.4 (6)	176.2 (1.1)		175.3 (1.3)	
Angles (degree)	C—M—C	90.8 (3)	90.0 (3)	90.7 (0.5)		88.3 (0.5)	
			92.4 (3)	89.4 (0.5)		91.5 (0.5)	
			87.4 (3)	90.5 (0.5)		88.4 (0.5)	
			90.2 (3)	89.4 (0.5)		91.7 (0.5)	
Déviations from middle plane (Å)	M	0.00	−0.024	−0.0001 (8)		−0.0002 (8)	
	C$_1$	−0.011	−0.006	−0.003 (12)		0.022 (14)	
	C$_2$	0.008	+0.003	−0.003 (13)		0.019 (12)	
	C$_3$	—	−0.020	0.018 (14)		0.006 (15)	
	C$_4$	—	0.000	0.016 (13)		0.006 (15)	
	N$_1$	0.007	−0.006	−0.032 (13)		0.061 (12)	
	N$_2$	−0.005	0.024	−0.065 (12)		0.147 (13)	
	N$_3$	—	0.003	0.042 (15)		−0.094 (15)	
	N$_4$	—	0.026	0.069 (13)		−0.038 (13)	
Angles between ⊥ to middle plane and the axis	a	—	—	81.03 (17)		93.83 (17)	
	b	—	—	80.60 (17)		84.71 (17)	
	c	4° 46′	6° 08′	18.72 (3)		10.76 (1)	
Distance M—M (Å)		3.364	3.387 (3)	3.733 (23)	3.743[b]) (24)		
Rotation angle between to planes		45°	26° 56′	0°			

[a]) Numbers in parantheses indicate standard deviations
[b]) Distance corrected for thermal motion
[c]) Estimate and not σ

the parameters of M(CN)$_4$ groups and their arrangement in columns. It can be noted that significant divergences from the D$_{4h}$ symmetry are extremely rare for the majority of the compounds.

Rb$_2$Pd(CN)$_4$, 1 H$_2$O		CaPd(CN)$_4$, 5 H$_2$O	KNaPt(CN)$_4$, 3 H$_2$O	[(C$_2$H$_5$)$_2$NH$_2$]$_2$ Pd(CN)$_4$	[(C$_2$H$_5$)$_3$NH]$_2$ Pd(CN)$_4$, 2 H$_2$O	[(C$_2$H$_5$)$_3$NH]$_2$ Pt(CN)$_4$, 2 H$_2$O
(7)		(5)	(12)	(10)	(10)	(10)
1.967 (20)	1.980[b] (20)	1.92 (5)[c]	2.02	1.970 (15)	2.05 (4)	2.07 (6)
2.038 (23)	2.050[b] (23)	2.00 (5)[c]	2.01	2.001 (12)		
—	—	1.99 (5)[c]	—	—		
—	—	1.94 (5)[c]	—	—		
1.164 (26)	1.166[b] (26)	1.12 (5)[c]	1.11	1.173 (16)	1.13 (5)	1.10 (9)
1.093 (28)	1.095[b] (28)	1.12 (5)[c]	1.12	1.156 (13)		
—	—	1.16 (5)[c]	—			
—	—	1.11 (5)[c]	—			
178.59 (2.27)		176 (4)[c]	177°	177° 51 (1.21)	175° 7 (2° 7)	180 (2)
176.74 (2.42)		172 (4)[c]	175° 30′	179°10 (0.95)		
—		168 (4)[c]	—			
—		176 (4)[c]				
90° 80′ (1.00)		90	86° 30′	89° 44 (0,44)	88° 7 ±2	91 (2)
89° 20′ (1.00)		92	93° 30′	—		
—		85	—			
—		93	—			
0.0		0.00		0	0	
0.0046 (0.0202)		0.003		−0.0201 (113)	−0.033 (21)	
0.0282 (0.0245)		0.03		−0.0011 (114)	−0.033 (21)	
−0.0046 (0.0202)		0.03		—	0.033 (21)	
−0.0282 (0.0245)		0.003		—	0.033 (21)	
−0.0028 (0.0193)		—		0.0126 (131)	0.022 (21)	
−0.0184 (0.0215)		—		0.0008 (106)	0.022 (21)	
0.0028 (0.0193)		—		—	−0.022 (21)	
0.0184 (0.0215)		—		—	−0.022 (21)	
—		87°	92°	—		
—		84°	84°	—		
5.89 (0.69)		6°	6° 30′	≃45°	≃42°	
3.720 (10)	3.730[b] (10)	3.42 (1)	3.25	6.35	8.22	8.17
17°		30°	36°	0°	0°	0°

171

3. Absorption Spectra of Tetracyanides

Figures 1—3 reproduce the absorption spectra of the three ions $Pt(CN)_4^{--}$, $Pd(CN)_4^{--}$, $Ni(CN)_4^{--}$ in the "free" state, that is, for compounds dissociated in aqueous solution.

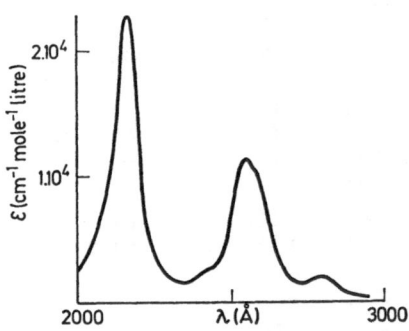

Fig. 1. Free ion $Pt(CN)_4^{--}$

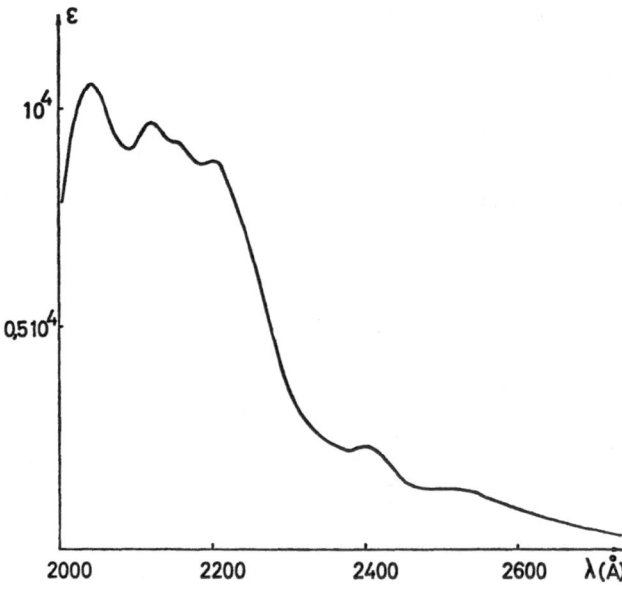

Fig. 2. Free ion $Pd(CN)_4^{--}$

With regard to the crystallized compounds, a distinction must be made on the one hand between the quantitative results in polarized light, which are limited toward the ultraviolet because of the very intense absorbance of the crystals, and on the other hand semi-quantitative results obtained in the ultraviolet with natural light from thin poly-crystalline films evaporated on a quartz plate.

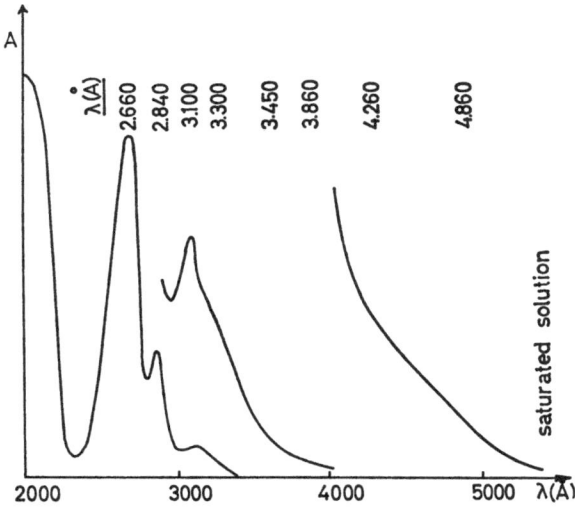

Fig. 3. Free ion $Ni(CN)_4^{--}$

It is pointless to show the numerous results obtained for each family of complexes since a survey of the whole of the spectra leads to the same conclusions. A curve representative of results obtained in our laboratory and by other researchers (13—19) using different techniques is given in Fig. 4, 5 and 6.

The results of classifying the absorption spectra are indicated in Tables 3 and 4.

1. The occurrence should first be noted of new intense bands in the visible for platinocyanides and in the near U.V. for palladocyanides. Although the platinocyanides are differently coloured while all the palla-docyanides are colourless, the absorption zone responsible for the colour of the platinocyanides would be analogous with the palladocyanides in the region of the shortest wavelengths.

173

The proximity of complexes stacked along the c axis has an equal effect on the two families of compounds. Fig. 7 demonstrates this analogy.

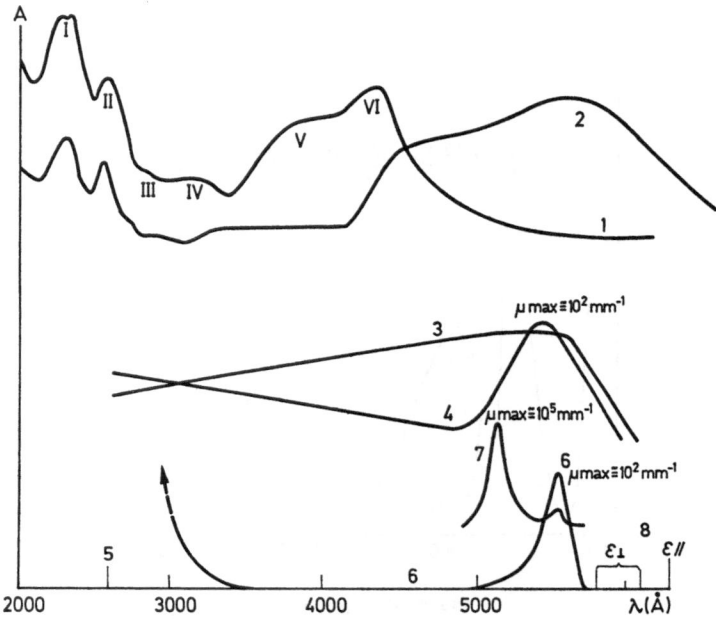

Fig. 4. 1. Absorption of microcrystals of $MgPt(CN)_4$, $2\,H_2O$; 2. Absorption of microcrystals of $MgPt(CN)_4$, $7\,H_2O$; 3. $\mu_\perp\,Yamada$ (microscopic techniques) for $MgPt(CN)_4$, $7\,H_2O$; 4. $\mu_{//}\,Yamada$ (microscopic techniques) for $MgPt(CN)_4$, $7\,H_2O$; 5. $\mu_{//}\,Moncuit$ (reflexion); 6. $\mu_{//}\,Moncuit$ (transmission); 7. $\mu_\perp\,Moncuit$ (reflexion); 8. emission measured by $Moncuit$

2. It is highly probable that the sharpest bands in the short wavelengths are due to the presence in the U.V. of intense bands of "free" ions with a weak shift towards the lowest energies. The shift depends very little on the nature of the ion. The first band of platinocyanides is clearly structured as two components, whereas the second band appears to be single. The free ion spectrum indicated the contrary.

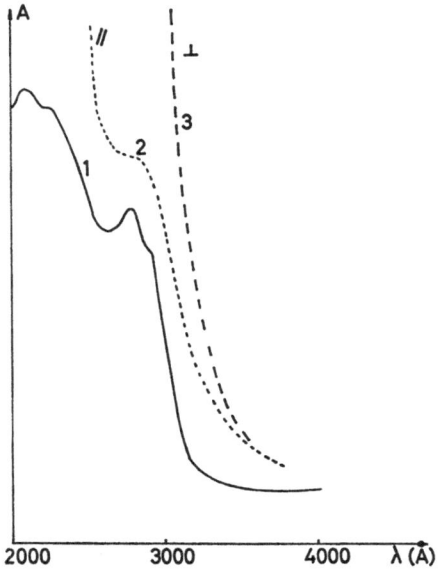

Fig. 5. 1. Absorption of microcrystals of SrPd(CN)$_4$, 5 H$_2$O; 2. $\mu_{//}$ (transmission) *Moncuit*; 3. $\mu_\perp$ (transmission) *Moncuit*

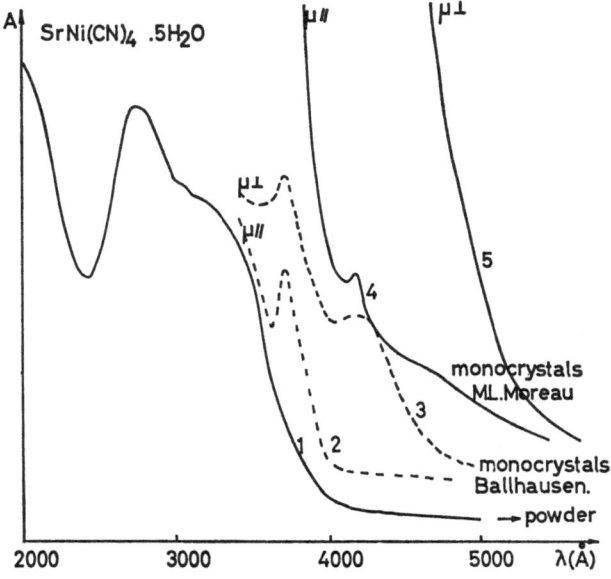

Fig. 6. 1. Absorption of microcrystals of SrNi(CN)$_4$, 5 H$_2$O; 2. $\mu_{//}$ on monocrystals (*Ballhausen*); 3. $\mu_\perp$ on monocrystals; 4. $\mu_{//}$ monocrystals (*M. L. Moreau*); 5. $\mu_\perp$ monocrystals

175

Table 3

Compound	Pt—Pt	band n° 1 very intense	Shoulder
MgPt(CN)$_4$, 7 H$_2$O	3.13	2,300 Å 43,480 cm^{-1}	
Y$_2$[Pt(CN)$_4$]$_3$, 21 H$_2$O	3.16	2,240 Å 44,640 cm^{-1}	2,340 Å 42,735 cm^{-1}
Ba, Pt(CN)$_4$, 4 H$_2$O	3.27	2,250 Å 44,444 cm^{-1}	2,350 Å 42,550 cm^{-1}
MgPt(CN)$_4$, 2 H$_2$O	3.36	2,250 Å 44,444 cm^{-1}	2,350 Å 42,550 cm^{-1}
CaPt(CN)$_4$, 5 H$_2$O	3.36	2,300 Å 43,480 cm^{-1}	
SrPt(CN)$_4$, 5 H$_2$O	3.57	2,260 Å 44,250 cm^{-1}	2,340 Å 42,735 cm^{-1}
Na$_2$Pt(CN)$_4$, 3 H$_2$O	3.67	2,220 Å 45,920 cm^{-1}	2,330 Å 42,920 cm^{-1}
Rb$_2$Pt(CN)$_4$, H$_2$O	3.69	2,230 Å 44,840 cm^{-1}	2,360 Å 42,370 cm^{-1}
		band n° 1 very intense	shoulder
[(C$_2$H$_5$)$_2$NH$_2$]$_2$ Pt(CN)$_4$ and solution	6.35	2,150 Å 46,510 cm^{-1}	2,200 Å 45,450 cm^{-1}

3. Taking into account the fact that the absorption spectrum in light polarized parallel to planar groups contains only a very weak absorption band in the visible range, it must be concluded that all the new intense bands of the crystalline state have a characteristic polarization perpendicular to the planes which produces the dichroism.

4. Nickel cyanides seem to behave differently from the isomorphous complexes of Pd and Pt. Already in the free state the spectrum indicates the presence of very weak bands due to d—d transitions, giving rise to the orange coloration of concentrated solutions. The passage to the crystalline state results in an absorption spectrum which is not very sensitive to the nature of the cation, except perhaps for BaNi(CN)$_4$, 4 H$_2$O, and is structured in the same way as the spectrum of the free ion except for the relative intensity of the absorption bands.

band n° 2 intense	band n° 3 very weak and shallow	band n° 4 weak and wide	band n° 5 intense without splitting	band n° 6 intense
2,550 Å 39,215 cm^{-1}	2,750 Å 36,360 cm^{-1}	—	4,800 Å 20,830 cm^{-1}	5,550 Å 18,020 cm^{-1}
2,570 Å 38,910 cm^{-1}	—	3,800 Å 26,315 cm^{-1}	4,800 Å 20,830 cm^{-1}	5,150 Å 19,420 cm^{-1}
2,580 Å 38,760 cm	—	3,300 Å 30,300 cm^{-1}	4,000 Å 25,000 cm^{-1}	4,560 Å 21,930 cm^{-1}
2,580 Å 38,760 cm^{-1}	2,850 Å 35,090 cm^{-1}	3,100 Å 32,260 cm^{-1}	3,850 Å 25,975 cm^{-1}	4,350 Å 22,990 cm^{-1}
2,600 Å 38,460 cm^{-1}	2,850 Å 35,090 cm^{-1}	3,150 Å 31,750 cm^{-1}	3,750 Å 26,670 cm^{-1}	4,340 Å 22,990 cm^{-1}
2,600 Å 38,460 cm^{-1}	—	—	3,350 Å 29,850 cm^{-1}	3,680 Å 27,170 cm^{-1}
2,580 Å 38,760 cm^{-1}	—	—	3,050 Å 32,790 cm^{-1}	3,360 Å 29,760 cm^{-1}
2,620 Å 38,170 cm^{-1}	2,840 Å 35,210 cm^{-1}	—	3,200 Å 31,250 cm^{-1}	3,575 Å 27,970 cm^{-1}

band n° 2 intense	shoulder weak	band n° 3 weak
2,550 Å 39,215 cm^{-1}	2,400 Å 41,670 cm^{-1}	2,800 Å 35,710 cm^{-1}

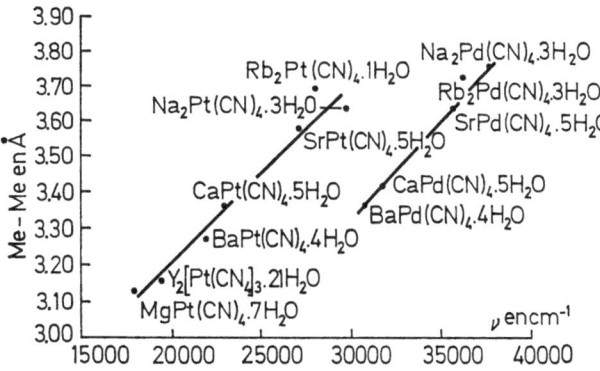

Fig. 7. Analogy between absorption bands of pallado- and platinocyanides

Table 4

	Maximum at highest wavelength	$Pd(CN)_4 - Pd(CN)_4$
$BaPd(CN)_4, 4 H_2O$	3,250 Å (30,770 cm^{-1})	3.36 Å
$CaPd(CN)_4, 5 H_2O$	3,150 Å (31,750)	3.42 Å
$SrPd(CN)_4, 5 H_2O$	2,800 Å (35,715)	3.63 Å
$Rb_2Pd (CN)_4, 3 H_2O$	2,760 Å (36,230)	3.72 Å
$Na_2Pd(CN)_4, 3 H_2O$	2,650 Å (37,735)	3.74 Å

4. Photoluminescence Spectra of Tetracyanides of Platinum

The luminescence of the platinocyanides has been measured by reflection in polarized light with the aid of a recording spectrofluorimeter. Emission is more intense when the vector of incident wave is parallel to the planar groups. The emission maximum (independent of the exciting wavelength) occurs in approximately the same place whatever the polarization of the incident wave (except for $MgPt(CN)_4 7 H_2O$) but there often appears a hypsochromic shoulder in the fluorescence spectrum related to the luminous vector perpendicular to the planar groups; this shoulder we can compare with the second band of $\mu_\perp$ spectra (Fig. 8).

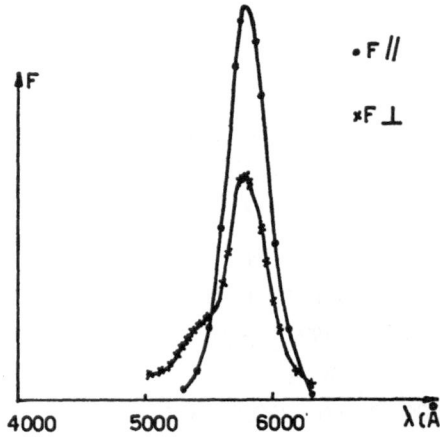

Fig. 8. Fluorescence of orange platinocyanides of lithium

The remarkable relationship between emission and absorption is apparent from Table 5.

Table 5

Compound	max. emission		max. absorption		M—M
	Å	cm⁻¹	Å	cm⁻¹	Å
$Y_2(Pt(CN)_4)_3$, 21 H_2O	6,100	16,400	5,150	19,420	3.16
$Li_2Pt(CN)_4$, 2 H_2O	5,780	17,300	4,900	20,400	3.18
$KNaPt(CN)_4$, 3 H_2O	5,600	17,860	4,800	20,830	3.25
$(NH_4)_2Pt(CN)_4$, xH_2O	5,300	18,870	4,700	21,280	3.26
$BaPt(CN)_4$, 4 H_2O	5,225	19,140	4,560	21,930	3.27
$CaPt(CN)_4$, 5 H_2O	4,900	20,400	4,340	22,990	3.36
$Rb_2Pt(CN)_4$, 1.5 H_2O	4,750	21,050	4,100	24,390	3.40
$K_2Pt(CN)_4$, 3 H_2O	4,520	22,120	4,050	24,690	3.50
$SrPt(CN)_4$, 5 H_2O	4,340	23,040	3,680	27,170	3.57

5. Interpretation of Spectroscopic Studies of Tetracyanides in the U.V. and Visible Range

I. Historical

The physical and chemical properties of compounds of transition metals have been the subject of much experimental and theoretical work and continue to interest a large number of research workers.

The progress realized in the interpretation of spectra of $M(CN)_4^-$ complexes cannot be dissociated from the current of ideas that have guided the researchers in their studies, embracing not only the planar groups with ligand CN but also all the compounds of transition elements $(d^1 \ldots d^9)$ bonded with the usual ligands.

We have compiled the experimental results and conclusions reached by different authors concerning the spectroscopic behaviour of the double cyanides with platinum, palladium and nickel. These results are grouped arbitrarily in three parts, dealing respectively with platinocyanides, palladocyanides and nickel cyanides. Finally, there is a paragraph reviewing the literature on the fluorescence of platinocyanides.

α. Platinocyanides

A. "Free" $Pt(CN)_4^{--}$ ion

a) Experimental data. The spectra of all the platinocyanides in aqueous solution are identical and correspond to the spectrum of the isolated $Pt(CN)_4^{--}$ ion.

Some solvent effects have been observed by *P. Mason* and *H. B. Gray* (*21*) but these effects are weak, as shown by Table 6.

Table 6

H₂O		E P A[a]) (300 °K)		E P A[a]) (77 °K)	
λ(Å)	ε(cm⁻¹ mole⁻¹ litre)	λ	ε	λ	ε
2,165	18,700	2,180	28,000	2,160	32,200
2,420	1,300	2,440	2,470	2,415	2,740
				2,550	15,460
2,560	9,000	2,575	13,420	2,600	11,950
2,800	1,270	2,800	1,800	2,810	2,430

[a]) 5:5:2 mixture of ether, isopentane and ethanol.

The resolution of the band at 2,560 Å into two bands (2,550 Å and 2,600 Å) is clearly established by the experiments of *R. Mason* at low temperatures. Measurements made by *Jørgensen* (*22*), *Ryskin et al.* (*23*), *Moncuit* (*24*) all agree on the existence of these five intense bands of the free ion in aqueous solution in the region between 2,000 and 7,000 Å.

b) Interpretation. The high intensity of these bands immediately led *Jørgensen* and the other authors cited to identify these bands with transitions of charge transfer, leaving *d* orbitals predominantly metallic towards combinations of empty orbital π of ligands.

The hypothesis of transition L (π_b of ligands) towards M has been rejected owing to the fact that their energy is higher in the isoelectronic $Au(CN)_4^-$ containing the more oxidizing central atom gold (III) (*27*).

The schemas of energy levels proposed by various groups of researchers differ, especially with respect to the position of the d_{z^2} (a_{1g}) orbital.

The order of the higher levels $\pi(*)$ (e_u, a_{2u}) . . . depends also on the author's views concerning the relative participation of the *p* orbitals in linear combinations with the orbitals of the ligands.

Table 7 is a summary of the whole of the data proposed by the authors cited above.

Table 7

$\lambda(\text{Å})$	oscillator strength (*Moncuit*)	*Gray* (25) *Ball-hausen* (1962)	*Ryskin et al.* (1963)	*Moncuit* (1965)	*Mason* and *Gray* (1968)
2,150	0,24		$d_{xz} \rightarrow p_z(a_{2u})$	$b_{2g} \rightarrow e_u^*$	$^1A_{1g} \rightarrow d^1E_u$
2,400	$\simeq 10^{-2}$	$^1A_{2u}$	$d_{xy} \rightarrow \pi^*(e_u, a_{2u})$	$a_{1g} \rightarrow e_u^*$	$^1A_{1g} \rightarrow {}^1A_{2u}$
2,550 ⎱ 2,580 ⎰	10^{-1}	^1E_u	$d_{z2} \rightarrow p_z(a_{2u})$	$e_g \rightarrow a_{2u}^*$	$^1A_{1g} \rightarrow c^1E_u$
2,800	$6,10^{-3}$	$^1B_{1u}$	$d_{z^2} \rightarrow \pi^*(e_u, a_{2u})$	$a_{1g} \rightarrow a_{2u}^*$	$^1A_{1g} \rightarrow {}^1B_{1u}$

B. Pt(CN)$_4^-$ Ion in a Crystalline Structure

In 1951 *Yamada* (*14*) studied by transmission monocrystals of platino-cyanides of magnesium, calcium and barium.

Using the microscopic method, he was able to state the position of the sharp absorption band $\mu_{//}$ (// to planar groups) at the following wavelengths:

MgPt(CN)$_4$, 7 H$_2$O	5,550 Å
CaPt(CN)$_4$, 5 H$_2$O	4,370 Å
BaPt(CN)$_4$, 4 H$_2$O	5,450 Å

The absorption bands $\mu_\perp$ appear very wide and have an intensity that can be compared to the spectrum of $\mu_{//}$, a fact which is not confirmed in any subsequent measurement.

Yamada was the first to point out the relation between the position of the absorption band and the parameter $\frac{c}{2}$ equal to the distance between the planes of Pt(CN)$_4^-$. He proposed to interpret these spectroscopic effects with the hypothesis of an axial interaction between stacked complex ions. According to *Yamada*, this proposed interaction would be even more remarkable than in *Magnus'* salt (Pt(NH$_3$)$_4$PtCl$_4$) where the opposite sign of charges of stacked groups electrostatically favours this type of arrangement.

In 1962 *Moncuit et al.* (*15*), working by reflection on platinocyanides of Mg, Ca, Sr, Ba, revealed in the visible a very intense absorption band

181

$\mu_\perp$, whereas in the spectrum of $\mu_{//}$ the reflection factor is practically constant in all of the visible.

The $\mu_\perp$ bands are located at the following wavelengths (Table 8):

Table 8

Compounds	$\mu_\perp$(Å)	$\mu_{//}$(Å)
MgPt(CN)$_4$, 7 H$_2$O	5,550 and 5,150	2,580
BaPt(CN)$_4$, 4 H$_2$O	4,400	2,630
CaPt(CN)$_4$, 5 H$_2$O	4,180	2,640
SrPt(CN)$_4$, 5 H$_2$O	3,530	2,670

The authors note that the effect of the distance Pt—Pt is important and propose the assignment $5d_{z2} \rightarrow 6p_z$ for the absorption bands $\mu_\perp$ (to the planes).

They suggest that the electrostatic field of the neighbouring ions may stabilize the d_{z2} and p_z orbitals, but draw attention to the fact that the effect of the crystallization will be even more important than the pressure effects obtained in Ptdmg$_2$ (dimethylglyoxime) where a shift of the band of 7,700 cm^{-1} is observed when P varies from 1 to 63,000 atm.

In 1963, *Ryskin et al. (16)*, working at 4.2 °K, found that the absorption band of the visible divides into two bands, the more intense, α being located at the longest wavelengths (Table 9).

Table 9

Compound	$\lambda\alpha$(Å)	$\lambda\beta$(Å)
BaPt(CN)$_4$, 4 H$_2$O	4,930	3,940
Li$_2$Pt(CN)$_4$, xH$_2$O	4,850	4,480
Yb$_2$Pt(CN)$_4$, 21 H$_2$O	5,950	5,100
Er$_2$Pt(CN)$_4$, 21 H$_2$O	5,920	5,200

These authors identified the α band as the transition $5d_{z2} \rightarrow 6p_z$ and for the β band proposed the alternatives $eg \rightarrow 6p_z$ or $b_{2g} \rightarrow 6p_z$.

Like the previously mentioned authors, they assumed that the field of axial deformation is able to stabilize the orbitals oriented along the z axis.

Moncuit (*15*) in 1964 completed the research on platinocyanides by studying the transmission of the same crystals and determined the position of the weak band $\mu_{//}$ in the visible at the following wavelengths:

$MgPt(CN)_4$, 7 H_2O	5,520 Å
$CaPt(CN)_4$, 5 H_2O	4,340 Å
$SrPt(CN)_4$, 5 H_2O	3,710 Å

Taking into account the relationship of these spectra with strong bands $\mu_{\perp}$, *Moncuit* was able to propose the attribution $5d_{z^2} \rightarrow 6p_z$ (A_{2u}) activated in the plane of the complex ion by a vibronic perturbation mechanism.

β) Palladocyanides

A. "Free" Ion $Pd(CN)_4^{--}$

a) Experimental data. The spectrum of the isolated $Pd(CN)_4^{--}$ ion has been studied simultaneously with the spectrum of the $Pt(CN)_4^{--}$ ion. It is less easy to index because of the presence of all the charge transfer bands in a limited wavelength region. *R. Mason* and *H. B. Gray* (*21*) seemed to find weak solvent effects and placed the bands at the wavelengths indicated in Table 10.

Table 10

H_2O		E.P.A. (300 °K)		E.P.A. (77 °K)	
λÅ	ε(cm^{-1} mole^{-1} litre)	λ	ε	λ	ε
2,040	10,400				
2,120	8,400			2,165	9,200
2,200	6,800	2,220	7,900	⎰ 2,220	9,800
				⎱ 2,260	5,700
				2,330	1,200
2,400	1,100	2,400	1,340	2,410	1,260

The presence of six bands was not noted by *Gray* and *Ballhausen* who recorded only the bands at 2,120, 2,200 and 2,400 Å. However, *Moncuit* indexed the spectrum of $Pd(CN)_4^{--}$ in aqueous solution and also indicated six absorption bands.

b) Interpretation. The identification of the spectrum of $Pt(CN)_4^{--}$ serves as the basis for the interpretation of the spectrum of $Pd(CN)_4^{--}$. The following table 11 summarizes the principal propositions.

Table 11

λ (Å) Moncuit	oscillator strength	Gray and Ballhausen	Moncuit	Mason and Gray
2,030	0.10		$^1E_u(b_{2g} \rightarrow e_u)$	d^1E_u
2,120	0.05	^1E_u	$^1E_u(e_g \rightarrow b_{2u})$	$^1A_{2u}$
(2,150)			$^1A_{2u}(e_g \rightarrow e_u)$	
2,200	0.07	$^1A_{2u}$	$^1E_u(e_g \rightarrow a_{2u})$	c^1E_u
2,315	0.01		$^1E_u(a_{1g} \rightarrow e_u)$	^1B_u
2,400	0.006	$^1B_{1u}$	$^1A_{2u}(a_{1g} \rightarrow a_{2u})$	$^1A_{2g}$

Contrary to these authors. C. K. Jørgensen thinks that the charge transfer transitions are π(CN) to $d(x^2-y^2)$ of Pd and not d(Pd) $\rightarrow \pi^*$(CN).

B. The $Pd(CN)_4^{--}$ Ion Embedded in the Crystal

In 1965 Moncuit and Macadre (19) studied by transmission the spectra of palladocyanides of Ca, Sr and Ba, and discovered in the $\mu_{//}$ spectrum the presence of weak bands at the following wavelengths:

$$BaPd(CN)_4, 4\ H_2O \qquad 3,290\ \text{Å}$$
$$CaPd(CN)_4, 5\ H_2O \qquad 3,150\ \text{Å}$$
$$SrPd(CN)_4, 5\ H_2O \qquad 2,800\ \text{Å}$$

The rest of the spectrum in the U. V. remains unexplored because of the excessive absorbance of the crystals.

Extending these measurements to $CaPd(CN)_4, 5\ H_2O$ in 1968, Moncuit and Lebras (20) were able to indicate the position of 2 bands // to the plane of the complex ion, one at 3,150 Å // b and the other // at a and b at 3,230 Å. Owing to the symmetry of the crystals, the authors were able to interpret these bands following an interaction model of individual moments of transition of $4d_{z2} \rightarrow 5p_z$ complex ions.

γ) The Nickel Cyanides

A. "Free" Ion $Ni(CN)_4^{--}$

Unlike the colourless solutions of $Pt(CN)_4^{--}$ and $Pd(CN)_4^{--}$, the $Ni(CN)_4^{--}$ ion is orange in concentrated solutions due to the presence in the visible of weak absorption bands.

Indeed, the electron transfer bands situated in the U. V. are shouldered on the long wavelength side by a large zone of weak absorption which is differently resolved according to different authors.

Table 12

λ (Å)	ε mole^{-1} cm^{-1} litre	Perumareddi (27) (1963)
2,660	16,150	Ch. tr.
2,840	4 820	Ch. tr.
3,095	770	$^1A_{1g} \rightarrow {}^1E_g$
3,300	409	$\rightarrow {}^1A_{2g}$
3,570	79	$\rightarrow {}^1B_{1g}$
3,860	7.8	$\rightarrow {}^3A_{2g}$
4,260	2	$\rightarrow {}^3E_g$
4,860	0.75	$\rightarrow {}^3B_{1g}$

Table 13

water		E.P.A. (77 °K)		Mason and Gray (1968)
λ (Å)	ε	λ (Å)	ε	
1,980	23,000			$^1A_{1g} \rightarrow d^1E_u$
2,680	12,000	2,685	15,230 $\Big\}$	$\rightarrow c^1E_u$
		2,760	5,230	
2,860	4,470	2,870	6,233	$\rightarrow {}^1A_{2u}$
3,100	703	3,110	838	$\rightarrow {}^1B_{1u}$
3,280	434	3,200	530	$\rightarrow {}^1A_{2g}$
4,440	2			$\rightarrow {}^3A_{2g}$

B. Monocrystals of Nickel Cyanides

In 1964, C. J. Ballhausen et al. (18) studied in polarized light the spectra of nickel cyanides of potassium, sodium, calcium, strontium and barium and found for all the compounds identical spectra containing a very weak band at 5,000 Å, a large band parallel to c at 4,350 Å and a sharp band both // and ⊥ at c at 3,700 Å.

Moreover, the nickel cyanide of barium has a weak band ⊥ (c) at 4,500 Å. The authors interpreted these bands by supposing that the excited state corresponding to the promotion of electrons to the $d_{x^2-y^2}$ orbital distorts the D_{4h} planar symmetry in favour of the D_{2d} symmetry. Under these conditions, the band at 4,350 Å is attributed to the $^1A_{1g} \to {}^1B_2$ transition permitted $//$ at c, and the band at 3,700 Å is attributed to the $^1A_{1g} \to {}^1E_u$ transition permitted both $//$ and $\perp$ to c, the degenerescence of the E state being capable of being split by a distortion of the Jahn-Teller type.

J. P. Dahl et al. (*28*) identified the unknown band of $BaNi(CN)_4$, 4 H_2O as magnetic dipolar transition $^1A_{1g} \to {}^1A_{2g}$ ($b_{2g} \to b_{1g}$). This band should appear for the other complexes $Ni(CN)_4^{--}$ but it would be hidden by the neighbouring bands which are more intense.

δ) The Luminescence of the Platinocyanides

As early as 1908, *Levy* carried out a qualitative study and classified the fluorescent platinocyanides according to the position of maximum emission. *Levy* tried to explain the shift of the fluorescence, relying only on the nature of ions. Considering only the alkali metal cations, he noted that the maximum of emission moved towards the red with a decrease in atomic weight, but for the whole class of compounds, he noticed no regular progression depending on the cation or degree of hydration.

In 1961, *Moncuit* and *Poulet* (*15*) studied the optical constants of platinocyanides of barium, calcium and magnesium and placed the maximum of emission $\varepsilon_{//}$ and $\varepsilon_\perp$ respectively at the following wavelengths.

Table 14

	$MgPt(CN)_4$, 7 H_2O	$BaPt(CN)_4$, 4 H_2O	$CaPt(CN)_4$, 5 H_2O
$\varepsilon_{//}$	6,300 Å	5,200 Å	4,900 Å
$\varepsilon_/$	6,100 Å	5,180 Å	4,850 Å
	5,810 Å	5,000 Å	4,670 Å

The shift of the fluorescence spectrum towards high frequencies with an increase in the distance between the planes of the complex ions varies in the same way as the positions of the absorption bands.

More recently (1965) *Ryskin, Tkachuk* and *Tolstoi* (*30*) measured the relaxation time τ of a large number of platinocyanides and found τ to be of the order of 10^{-6} to 10^{-7} sec. They also noted that the independence of the luminescent spectrum with regard to the exciting radiation shows that the redistribution of the electrons on the excited levels responsible for the emission takes place very rapidly relative to the lifetime of the excited state.

In 1965 *Tkachuk* and *Tolstoi* (*31*) also studied the emission of frozen solutions in its dependence on the concentration and kinetics of the cooling process. At high concentrations ($> 10^{-3}$ molar), the emission is that of monocrystal hydrated to the maximum. In medium concentrations, emission shows several bands due to the partially hydrated state. At very low concentrations ($< 10^{-5}$ molar), the same emission is obtained for all the cations and is characterized by two bands (4,300 and 5,100 Å) due to the monomolecular ($H_2Pt(CN)_4$) and bimolecular ($H_2Pt(CN)_4)_2$ agregates.

No definite interpretation has been proposed for luminescence of platinocyanides; the possible existence of a triplet state has not yet been fully studied.

II. Discussion

This review shows that, despite all the investigations, the order of the energy levels of $M(CN)_4^-$ ions has not yet been well established, either for free ions or for ions inserted in the crystalline structure.

With regard to the platinocyanides, the existence in the visible spectrum of two intense absorption bands for the crystalline state is a problem which has not been entirely solved.

Our results on microcrystals clearly show the two absorption bands of unequal height and the characteristic polarization perpendicular to planes.

Yamada found a spectrum of $\mu_\perp$ probably containing several bands, but the appearance and the intensity of this spectrum has never been reproduced in any subsequent work.

Ryskin et al. have clearly resolved two bands at low temperatures but they have not stated the precise polarization.

Finally, *Moncuit* and *Poulet*, working by reflection, have measured only an absorption band related to the electric vector perpendicular to the planes, except for the platinocyanide of magnesium.

If the only experimental parameter which seems to determine directly the evolution of the optical properties is the proximity and the relative disposition of the planar groups, it is normal to ask whether there may be an interaction between these planes, bringing modifications inside the electronic schemes of the isolated groups. Several authors (see Introduction) have postulated such an interaction.

Some have tried to elucidate the matter by bringing in new experimental data. Thus, *Krogmann* (*2*) working on partially oxidized complexes (with Pt) found that the M—M distances were particularly contracted.

He explained the formation of an energy band d_{z^2} and p_z binding and mixing by configuration interaction.

Our studies carried out on compounds having a long distance between planes (compounds with bulky organic cations) and those having a very short distance ($K_2Pt(CN)_5$ and other cations) can be inserted in this picture, as well as the study of mixed compounds containing $Pt(CN)_4^-$ and $Pd(CN)_4^-$ in the same columns.

a) organic complexes with long M—M distance have an absorption spectrum which may be superposed on the spectrum of the "free" ion.

The intensification of the band identified by *Moncuit* $a_{1g} \rightarrow a_{2u}$ for the free ion (situated at 2,800 Å for the platinocyanides and at 2,400 Å for the palladocyanides) when working on thin polycrystalline sheets can easily be explained by the preferential orientation of the crystals on the quartz plate. It must be remembered that this band related to incident vibrations polarized perpendicular to planar groups no longer appears clearly at this energy for the complexes with alkali-metal and alkaline-earth cations.

We think that this band, which is characteristic of groups in the free state and of crystals with organic groups, is the final manifestation in the ultraviolet of one of the bands of series V or VI of the crystalline state (see Tables 3 and 4).

The attribution $nd_{z^2} \rightarrow (n + 1)p_z$ is the usual term proposed for one of the bands of the visible and does not upset the interpretations of the spectrum of the free ion, provided it is accepted that there is no important linear combination between metallic p_z orbitals and π^* (a_{2u}) of the ligands.

b) Complexes having short Pt—Pt distances (the analogous derivatives of Pd or Ni have not been synthesized) have an absorption spectrum in solution which may be superposed (intensity and position) on the spectrum of the free ion $Pt(CN)_4^-$. In the crystalline state the compounds (having a metallic brown aspect) strongly absorb in the ultraviolet, the visible and the near infrared. From spectra obtained by transmission, it is not possible to indicate the position of the absorption bands.

These compounds with short metal-metal distance look very similar to the compounds of *Krogmann* and it seems that the stratified chains of $Pt(CN)_4^-$ ions are stabilized by the same type of interaction.

In fact, these compounds are the only ones able to possess a real chemical bond between the adjacent groups. Indeed, the calculation of the potential energy of Van der Waals forces (*O. Dideberg*), taking into account only the squares of the CN ligands stacked without a central ion, shows that the minimum energy is required for the distance $\frac{c}{2}$ and angle γ (rotation angle of 2 planes) as measured experimentally in the complexes. This surprising result is valid for nearly all the tetra-cyanides within the limits of error imposed by the starting approximations, but $K_2Pt(CN)_5$ is found clearly outside this region.

Moreover, one should also note the great ease with which the usual complexes ($\frac{c}{2}$ going from 3.50 to 3.70 Å) are able to dehydrate (below 100 °C) and thereby to reorganize themselves in order to form a new structure where the c parameter is considerably modified in one way or another, as shown by the spectacular colour changes (*32*) (*33*).

$K_2Pt(CN)_5$, 3 H_2O does not possess this property. It loses its three water molecules at 120 °C and then rapidly decomposes.

c) The mixed derivatives of Pt and Pd present in the ultraviolet the superposition of the bands of free ions with the shift accompanying the engagement of the ions in the crystals.

The bands of the visible, extrapolated for suitable M—M distances, coincide with two new absorption components due to the intermediate configurations of platinum surrounded by two palladium and of palladium enclosed by two platinum.

The mixed compounds are fluorescent, with colours intermediate between those of palladocyanides and of platinocyanides, and at a low percentage of Pt two bands of emission appear clearly. The new band, which decreases very rapidly in intensity when the amount of Pt is increased, probably originates in the intermediate configurations where the $Pt(CN)_4^-$ ion is held between two groups of $Pd(CN)_4^-$. The quantitative results of this study on mixed compounds will be published later.

Acknowledgements. The author wishes to thank Prof. *Brasseur* and Prof. *Toussaint* for helpful discussion and Prof. *Jørgensen* for reading and criticizing the manuscript.

The work was made possible by a grant from F. N. R. S. (Fonds National de la Recherche Scientifique).

6. References

1. *Fontaine, F., Moreau, M. L., Simon, J.*: Bull. Soc. Franc. Miner. Crist. *91*, 400—402 (1968).
2. *Krogmann, K.*: Angew. Chem. Intern. Ed. Engl. *8* (1), 35 (1969).
3. *Krebs-Larsen, F., Grønbaek-Hazell, R., Rasmussen, S. E.*: Acta Chem. Scand. *23*, 61—69 (1969).
4. *Brasseur, H., de Rassenfosse, A.*: Mem. Acad. Roy. Belg., Cl. des Sci., vol. XVI (1937).
5. *Fontaine, F.*: Bull. Soc. Roy. Sci. Liege. 9/10, 437—444 (1968).
6. *Lambot, H.*: Bull. Soc. Roy. Sci. Liege. *12*, 522 (1943).
7. *Dupont, L.*: Bull. Soc. Roy. Sci. Liege. 9/10 (1969); Acta Cryst. B *26*, 964 (1970).
8. *Ledent, J.*: Dr. Thesis, Liege University.
9. *Dupont, L.*: Dr. Thesis, Liege University.

10. *Jerome, S.:* Dr. Thesis, Liege University.
11. *Holt, E. M., Watson, K.:* Acta Chem. Scand. *23*, 14—28 (1969).
12. *Moreau-Colin, M. L.:* Bull. Soc. Franc. Mineral. Crist. *91*, 332 (1968).
13. — Chim. Phys. *67*, n⁰ 3, 498 (1970).
14. *Yamada, S.:* Bull. Chem. Soc. Japan *24* (3), 125 (1951).
15. *Moncuit, C., Poulet, H.:* J. Phys. Radium *23*, 353 (1962). — *Moncuit, C.:* J. Phys. Radium *24*, 833 (1964).
16. *Ryskin, A. I., Tkachuk, A. M., Tolstoi, N. A.:* Opt. Spectry. (USSR) (Engl. Transl.) *17*, 724—727 (1964).
17. *Moreau-Colin, M. L.:* Bull. Soc. Franc. Mineral. Crist., *88*, 608—609 (1963).
18. *Ballhausen, C. J., Bjerrum, J. N., Dingle, R., Eriks, K., Hare, C. R.:* Inorg. Chem. *4* (4), 514 (1965).
19. *Macadre, A., Moncuit, C.:* Compt. Rend. Acad. Sci. Paris *261*, 2339—2342 (1965).
20. *Moreau-Colin, M. L.:* Bull. Soc. Roy. Sci. Liege. 11—12 (1965).
21. *Mason, W. R., Gray, H. B.:* J. Am. Chem. Soc. *90*, 5721 (1968).
22. *Jørgensen, C. K.:* Absorption spectra and chemical bonding in complexes. Oxford: Pergamon Press 1962.
23. *Ryskin, A., Tkachuk, A. M., Tolstoi, N. A.:* Opt. Spectry. (USSR) (Engl. Transl.) *17*, 565—570 (1964).
24. *Moncuit, C.:* Chim. Phys. *64*, n° 3, 494 (1967).
25. *Gray, H. B., Ballhausen, C. J.:* J. Am. Chem. Soc. *85*, 260 (1963).
26. *Lebras, R., Moncuit, C.:* Compt. Rend. Acad. Sci. Paris *267*, 1032 (1968).
27. *Perumareddi, J. R., Liehr, A., Adamson, A. W.:* J. Am. Chem. Soc. *85*, 249 (1963).
28. *Dahl, J. P., Dingle, R., Vala, M. T.:* Acta Chem. Scand. *23*, 47—55 (1969).
29. *Levy, L. A.:* J. Chem. Soc. *93* (2), 1447 (1908).
30. *Ryskin, A. I., Tkachuk, A. M., Tolstai, N. A.:* Opt. Spectry (USSR) (Engl. Transl.) *21*, 31 (1966).
31. *Tkachuk, A. M., Tolstoi, N. A.:* Opt. Spectry (USSR) (Engl. Transl.) *20*, 570 (1966).
32. *Bergsøe, P.:* Acta Chem. Scand. *16*, 2061 (1962).
33. — *Hansen, P. G., Jacobsen, C. F.:* Nucl. Instr. Methods *17*, 325 (1962).

Received June 16, 1971

Structure and Bonding: Index Volume 1-10